BOILER FEED WATER TREATMENT

BOILER FEED WATER TREATMENT

by

F. J. MATTHEWS

1951
CHEMICAL PUBLISHING CO. INC.
NEW YORK., N.Y.

Boiler Feed Water Treatment

ISBN: 978-0-8206-0098-7

Chemical Publishing Company:
www.chemical-publishing.com
www.chemicalpublishing.net

First Edition:

© **Chemical Publishing Company, Inc.** - New York 1951

Second Impression:

Chemical Publishing Company, Inc. - 2011

Printed in the United States of America

PREFACE TO FIRST EDITION

ONE OF THE MOST IMPORTANT ASPECTS OF EFFICIENT STEAM GENERATION is the question of treatment of the boiler feed water. If this is incorrectly treated, or perhaps not treated at all, a crop of operating troubles is the result. A considerable amount of work has been carried out on this subject, particularly in the last ten years, and new methods of treatment have been developed as well as modifications of older methods. This has been rendered imperative by the great changes in boiler design which have taken place in recent years, but the result is a somewhat mixed blessing. While the rapid advance has proved beneficial to operators of large installations, super-pressure plants, etc., it has only confused the small "ordinary" plant operator with a wide variety of treatments. This is the more unfortunate since the large super-pressure plants have their own chemist or have sufficient financial resources to refer their problems to a boiler water chemist, whereas the small operating engineer usually has to worry out his problems for himself. It is hoped this little book will help to correct such a situation, to collect in a coherent form information on the subject, much of which information has so far appeared only in the form of pamphlets, articles, and research reports. With this object in view the book has been divided into sections dealing with the principal operating problems separately. It is not suggested by this that problems are in separate and individual watertight compartments, but this seemed a convenient way to emphasize the primary object of a particular treatment or procedure. Water softening, of course, has the two-fold object of preventing scale and corrosion.

Much has already appeared in the technical Press in the form of articles by myself, and I have to acknowledge the kind permission of numbers of Editors to use any portions of such articles as may have seemed desirable. In particular, my thanks are due to the Editors of *Cheap Steam, Power Engineer, Engineering and Boiler House Review, Colliery Engineering, World Power, Brewers' Journal, British Clayworker,* and *Power Laundry.* I am also indebted to makers of plants and instruments for their permission to reproduce photographs of their outfits and for the loan of blocks for many of such illustrations.

F. J. MATTHEWS.

PREFACE TO THIRD EDITION

THE PRESENT CRITICAL FUEL SITUATION GREATLY ENHANCES THE importance of efficient power plant water treatment, since substantial fuel savings can be effected in various directions if the raw water supply is correctly treated before passing to the boiler plant. The steadily continuing progress in the development of new and improved treatment techniques and plant further emphasizes the need for works' managements and engineers to keep informed on the subject.

In this third edition the text has been completely revised, and rewritten where necessary, to cover certain notable developments since the last edition and to bring the text on scale formation and corrosion in particular up to date in relation to current research conclusions. New matter on a number of aspects of the subject has also been added to this edition in response to various requests from correspondents.

Again I am indebted to a number of firms for the loan of blocks to illustrate their plants or instruments, and my thanks are also due to a number of correspondents for drawing attention to advisable improvements which have been incorporated.

F. J. MATTHEWS.

CONTENTS

LIST OF HALF-TONE ILLUSTRATIONS

Also 41 line diagrams in text

PART ONE

NATURAL WATER SUPPLIES

THE ORIGINAL SOURCE OF ALL NATURAL WATER SUPPLIES IS RAIN, IMpurities being collected in the passage through the air and the various strata of the earth as the water passes through them. Although waters differ widely in composition according to their previous "life" history, the impurities collected during this journey to the consumer are comparatively few. The mineral salts consist chiefly of the bicarbonates, carbonates, chlorides, sulphates, and nitrates of calcium, magnesium, and sodium ; small amounts of silica may be present and sometimes a water contains appreciable quantities of iron and aluminium salts. In addition to the mineral constituents there are usually some dissolved gases, such as oxygen or carbon dioxide and possibly acid vapours. Finally, the water may contain suspended matter and this may be of either mineral or organic origin. The relative amounts of these impurities will depend upon the origin of the water supply, and as the suitability of a given water for boiler feed purposes depends on the effects of the impurities it will be advisable to consider briefly the various types of natural water supplies.

Rain Water

Rain water will, of course, take up impurities it meets in falling through the atmosphere. These are dissolved gases and dust. It is usually considered a soft water and is certainly more pure than most natural waters. It does, however, contain substantial amounts of dissolved oxygen and carbon dioxide, and these are corrosive gases. A more dangerous source of corrosion in rain water is the acidity which it may acquire in passing through the atmosphere. Modern industrial life results in the production of acid gases in the atmosphere and this acidity will vary according to the locality. Thus, in towns and manufacturing areas where large amounts of coal are burnt, the sulphuric acid content of the rain is much higher than in the country. Another acid, nitric acid, is produced during thunderstorms by the electrical oxidation of nitrogen in the atmosphere. The dust dissolved by rain in passing through the air may be of a very varied composition. In industrial areas it may contain iron oxide, silica, sulphate and carbonate of lime, ammonium chloride, magnesium sulphate, sodium chloride, and organic matter. Rain falling in coastal areas may contain appreciable amounts of sodium chloride.

If rain water is used as a source of supply, therefore, close watch has to be kept that no free acidity or dissolved gases such as oxygen pass to the boiler. Even though the amount of scale-forming salts may be small, under some conditions it may lead to trouble, while some of the hardness may be due to unstable salts, which will break down in the boiler to give acids. Although rain water is considered soft, it is advisable to treat it and prevent the development of these troubles.

Surface Water

When rain water reaches the ground it can either pass through the earth or remain on or near the surface and form streams, lakes, or rivers. In either case the acids and acid gases it contains assist the solution of further impurities.

Surface waters are usually saturated with dissolved gases, since they are continually exposed to the atmosphere. They also contain appreciable amounts of suspended matter. In moorland districts they may become contaminated with organic or peaty matter, which is liable to decompose and give humic acids. The water, although very little harder than rain water, is still corrosive and requires treatment.

River Water

Rivers are mainly fed by surface water and the character of the water therefore depends largely on the nature of the collecting grounds from which they are fed. As a rule they are less pure than rain water supplies. The composition is subject to considerable variation, since the amount of dissolved solids will vary with the rate of flow of the river and also with the time of the year. Thus, in winter floods more silt and suspended matter will be carried down ; the organic matter will also be higher in such circumstances. The amount of suspended matter may be thought to vary according to the point at which samples are taken. This is true to a certain extent ; the suspended matter will be high at the source or where the rate of flow is rapid, and low when the river widens or the rate of flow diminishes, owing to gradual deposition of the suspended matter. It should be remembered, though, that small streams and other tributaries repeatedly add to the suspended matter, and accordingly river water is rarely free from suspended matter, which is not easy to remove owing to the fineness of division of the particles. River water also contains much dissolved oxygen and carbon dioxide. In addition, an important factor is the possible existence of factories on the banks of the river. Trade effluents, waste liquors, or sewage are frequently discharged into such rivers.

Spring Water

It has already been mentioned that rain water contains appreciable amounts of dissolved carbon dioxide. This makes the water acidic (quite apart from any other acid substances present) and adds to the solvent power. When the rain passes through the ground, therefore, it takes up calcium and magnesium carbonates in the form of bicarbonates. These substances decompose under the action of heat to give back the gas and the original carbonates. Spring water does not usually contain suspended matter owing to the filtration effect of passage through porous strata. The acidity of rain water caused by mineral acids is not found in spring water since it is neutralized by oxides in the earth. Neither is the dissolved oxygen content excessive because the water has not been continually exposed to the atmosphere. In the case of spring water, therefore, the aim is to remove whatever dissolved gas it contains and to treat the water to counteract the effects of the scale-forming and corrosive dissolved salts.

Well Water

Wells may be divided into two classes, shallow and deep, but only the latter are used for industrial purposes. Deep wells may be compared with springs, the difference being that springs are formed by the water's percolating the strata until it reaches a fault in an impervious layer or stratum, whereas in the case of wells the water accumulates at the impervious stratum until the well is tapped. The water is therefore similar to that of springs, the dissolved solids consisting of sulphates and carbonates. In the case of some deep wells the water is strongly alkaline, owing to sodium carbonate content. This is the result of zeolitic action, the water passing through natural beds of zeolite or aluminosilicate at some time or other and so having its calcium and magnesium carbonates changed to sodium carbonate by the base exchange action of the mineral. Wells have been known in which water drawn from one depth was hard and contained calcium and magnesium carbonates and sulphates, but when the well was sunk further the water became alkaline and soft.

Canal Water

Canal waters as a source of boiler feed are open to several objections. They usually contain excessive amounts of suspended matter, their composition varies, and they may contain acid substances. Factories sometimes dispose of effluents via canals, and these may contain very objectionable constituents. The water from mines is sometimes pumped into canals, and this is usually acid in character and in addition carries a good deal of dissolved sulphates and iron in suspension.

Pit Water

Water from the seams of many mines is strongly acidic and in addition contains substantial amounts of sulphates. These latter may be scale-forming sulphates or salts which decompose to give free sulphuric acid. The acidity varies according to the mine, those with more sulphur in the workings giving the most acid waters. Water pumped from old mine workings is particularly corrosive and objectionable. Sometimes two different waters may be obtained from the same pit, one seam giving a hard, corrosive water, and another giving an alkaline water. This is unusual, but where it does occur the two waters may possibly be combined to give a water suitable for boiler feed. As a rule, however, mine waters are not suitable for boiler feed purposes without heavy treatment.

Even then the amount of dissolved solids in the treated water may be excessive.

Effects of Drought

Industrial waters vary from time to time in their composition, and frequent analysis is necessary to be sure current treatment is satisfactory. These variations may be due to various influences. Well and spring waters are affected by rainfall and in times of drought the normal supplies may be diluted by water forced in from quite different sources. The character of the water may be completely changed due to the fact that normally rain supplies the bulk of the water and dilutes a more objectionable minor part of the well water obtained from some other source; then during drought periods the dilution will be smaller or perhaps even negligible, when the evil character of the second source will become apparent. Rivers and canals are similarly affected. The possibility of contamination at times by factory effluents has already been touched upon. If artificial fertilizers are put down on agricultural land drained by a given river, the acidity may increase. The passage of barges along canals may disturb sediment and other impurities. With regard to rivers, at times of drought there will be greater influx of sea water at tidal estuaries, and this will seriously affect the operation of softeners owing to the high quantity of dissolved and corrosive salts. Constant vigilance is therefore required whenever there is reason to suspect variation or possible variation in a source of water supply.

Condensate as Feed Water

The use of condensate from condensing engines and turbines as boiler feed is of great advantage with modern steam plant. Not only is the danger of scale formation considerably reduced, but such reduction is also accompanied by economy in water treatment. This is particularly important where the only available water supply forms very bad boiler feed without heavy treatment. The recovery of all condensate from machines using steam is, of course, impracticable owing to the losses incurred during power production, but it is advantageous to recover as fully as possible and return the water as boiler feed. The more condensate is used the less solids there are entering the boiler and consequently less blowing-down is necessary. There are, however, certain precautions necessary in the use of condensate, which are of considerable importance.

Condensate invariably contains gases in solution which, if permitted to enter the boiler and economizer, will lead to extensive corrosion. The usual slight acid reaction of condensate is generally considered to be due to carbonic acid or dissolved carbon dioxide gas, though Paul has suggested that it may be due to some organic acid formed by the decomposition of sodium carbonate in the boiler under high pressure conditions.

The other gas which causes trouble in condensate is dissolved oxygen, and it has been shown by Riviere that most of this gas in condensate is due to absorption from the air since he obtained the following figures for turbine condensate:

TABLE 1

Condensate at	Temperature °F	Dissolved Oxygen c.c. per litre	Saturation per cent.
(*a*) Lea Recorder inlet	70	0·136	1·6
(*b*) Lea Recorder outlet	70	2·14	26·2
(*c*) Feed heater	133	3·82	91·0
(*d*) Hot well	110	4·22	93·3

The writer also recalls some boiler tests in which, for purposes of checking its weight, the condensate was allowed to flow full-bore from a 1 in. pipe into an open drum; all concerned were surprised to find that in the course of filling the drum, the condensate picked up sufficient oxygen to show two-thirds saturation on testing. It is evident, therefore, that a little increased care in handling condensate will often reduce appreciably the treatment necessary to counteract its influence.

Another source of trouble is leakage of the condenser, since contamination by the circulating water may result in unstable corrosive salts and scale-forming minerals reaching the boiler. When a large amount of condensate is used, even a very small condenser leak may admit sufficient of these minerals to the feed to neutralize the slight alkalinity of the make-up. Cases have been known where the hardness of the make-up has been two or three parts per 100,000, but after the condensate was added the figure was almost doubled and the final feed slightly acid. Very small amounts of unstable minerals can cause substantial corrosion, and at high evaporation rates very little scaling matter will result in serious deposition, so that a careful watch for leakages is essential. There are two good methods of detecting condenser leaks. The chemical method is to determine the chloride ratio for circulating water and condensate in the same way as is done for boiler blow-down and boiler water (see page 183). Then the percentage of leakage will be given by $100\ c/C$ where c is the chloride content of the condensate and C is that for the circulating water. Where the chloride content of the circulating water is low, the leak may be detected electrically by an instrument known as the "Dionic" tester. This operates on the principle that there is a relationship between the purity of water and its electrical conductivity, the purest water having the lowest conductivity. Thus, with condensate a "Dionic" meter will show a certain conductivity, but if leakage is taking place the conductivity and therefore the meter reading will change.

Water Blending

An operation analogous to the use of condensate with softened feed but less commonly practised is the blending of two or more different waters. This may lead to substantial savings in the chemicals required for treatment. Such procedure is admittedly not always possible, but where two waters of a different type are available, blending is well worth consideration.

For the purpose of blending, waters can be considered in three classes, viz., acid, alkaline, and hard or neutral waters. The following may be considered as concrete examples of each of these types:

TABLE 2

	Alkaline	Hard or Neutral	Acid
Calcium carbonate	4·07	8·73	—
Magnesium carbonate	6·18	8·16	—
Calcium sulphate	—	20·03	28·10
Magnesium sulphate	—	15·63	20·15
Magnesium chloride	—	4·10	—
Sodium sulphate	14·61	28·58	37·80
Sodium chloride	23·40	15·36	—
Sodium carbonate	33·40	—	—
Sulphuric acid	—	—	7·60
Iron sulphate	—	—	31·30
Silica	0·16	0·27	—
	81·82	100·86	124·95
			grains per gallon

These are by no means the worst of each type and are merely selected as illustrative examples of the principles of blending. These principles can always be applied, but it must again be emphasized that the effect of blending depends entirely on the analyses and quantities of the waters used. The hard water is termed neutral merely for comparison; actually such waters may be either slightly acid or slightly alkaline.

The most obvious case for blending is the mixture of acid and alkaline waters. It will be evident at the outset that the free acid in the acid water may be neutralized by the sodium carbonate of the alkaline water and corrosive effects thus eliminated. It is equally evident that the correct amounts of each water, as determined by its analysis, must be used or else free acid may be left unneutralized in the blend. By suitable adjustment of the quantities of water a 100 per cent saving in chemicals required for neutralization of the acid may be effected. Similar economies can be effected by blending an alkaline water with a hard or neutral water, the sodium carbonate of the alkaline water serving to treat the permanent hardness. In the above examples, an equal blend of the two waters would result in a 98·5 per cent saving of soda ash for the hard water, and, of course, comparable economies if one considered the blending from the viewpoint of treatment of the alkaline water.

Apart from chemical economies, an important advantage is the substantial reduction in total dissolved solids passing to the boilers, and if condensate also is available for inclusion in the feed, the mineral content will be still further reduced. In fact, hitherto useless waters may be reduced sufficiently in salinity to be utilized. More savings will follow, too, from the lower rates of boiler blow-down required, reduced risk of priming and wet steam production, etc. It is essential, however, that strict analytical control should be maintained over the whole operation, otherwise results may be obtained worse than if no treatment of any description were given.

PART TWO

SCALE FORMATION

IT HAS BEEN SHOWN IN THE PRECEDING SECTION OF THIS BOOK THAT ALL natural waters contain some solid matter in solution and when they are evaporated to dryness this is left behind. Precipitation often occurs, however, long before complete dryness is reached, because the solution becomes saturated with one or more of its components, and when evaporation takes place in a boiler the result is a production of a sludge in the boiler water or a scale on the plates and tubes. The scale and sludge may consist of any salt present in the water ; for example, cases of sodium sulphate forming the major part of a scale are on record, but deposits of soluble salts of this kind are almost always unnecessary and are generally the result of gross inattention to the regulation of the concentration of the boiler water.

Whilst it is undesirable that either scale or sludge should be formed in a boiler, there can be no question but that scale is the more objectionable. The presence of sludge in large quantity may be a cause of wet steam and sometimes necessitates an excessive blow-down, but the troubles due to scale adherent on the heating surface are much more serious.

Effect of Scale

A layer of scale on a boiler plate or tube acts essentially as an insulator, reducing the ease with which heat is transferred from the outside of the metal to the water on the inside. The actual conductivities of different boiler scales vary considerably, but measurements by various workers have shown that a scale of anhydrite (calcium sulphate), dense and adherent to the metal surface, such as is produced from a water containing calcium sulphate in a high-pressure boiler, has a specific conductivity of about 1·0–1·5 B.Th.U. per sq. ft. per hr. per foot thickness per degree Fahrenheit. The average specific conductivity of steel is about 20 times as much, so that a layer of scale as thick as the wall of a water tube will increase the resistance of the tube to heat flow at least 20 times. Many scales have lower conductivities than this, and are correspondingly more objectionable.

The reduction thus brought about in the conductivity of the tube walls has two results. Firstly, less heat gets through and therefore there is a certain waste of fuel. Secondly, the decreased conduction of heat from the inside of the metal walls results in these walls becoming hotter. Both of these effects are, of course, always present, but their relative values vary greatly with conditions.

In boilers where the heat is transferred mainly by conduction between the hot gases and the tubes and boiler walls, a good deal of loss of heat may take place. A critical examination of this question has shown, however, that losses due to this cause are not so large as was formerly thought. Some recent values indicate that the heat losses to be expected under average conditions for a scale of 1/5 to 1/10 in. thick vary from $1\frac{1}{2}$ to 6 per cent.

Partridge and White,[1] of the University of Michigan, have investigated the thermal effects of boiler scale in a small electrically heated boiler operating at various pressure up to 150 lb. per sq. in. gauge. Valuable conclusions were reached by them. They found that the physical structure of a scale has a profound effect on its conductivity, but the chemical composition has only a slight effect, excepting in so far as it influences the physical structure. Thus, the heat conductivity of a thin porous scale is lower than that of a much thicker but continuous non-porous scale. This is explained by the fact that the interstices of the porous scale will be filled with steam and not with water, owing to the high temperature of the scale deposit compared with the water in its vicinity. It is this steam that is the real insulator and not the scale. They also point out that the resultant heat drop across a layer of scale may be calculated by means of the fundamental heat flow equation. This may be written in its simplest form as follows:

$T=\frac{QL}{K}$ where T is the temperature drop in deg. F., Q is the heat transferred in B.Th.U. per sq. ft. per hour, L is the thickness of scale in feet, and K is the coefficient of thermal conductivity of the scale in B.Th.U. per sq. ft. per foot thickness per deg. F. In modern boilers the heat input rates may be of the order of 70,000 to 100,000 B.Th.U. per sq. ft. per hour, and the safe maximum continuous temperature for boiler tubes of mild steel is about 900° F. Using these figures the limiting thickness of scale can be calculated from the above equation. Suppose the boiler pressure is 475 lb. per sq. in.; then the temperature of the water in the boiler is 465° F. Under these conditions, neglecting the temperature drop from the metal heating surface to water, the temperature of the metal is 435° F. below the safe maximum continuous temperature of 900° F. Taking a heat transfer rate of 70,000 B.Th.U. and an average conductivity of 1·5 (a fairly compact scale), the equation gives a limiting thickness of scale of 0·11 in. Forcing the boiler increases the heat transfer rate and therefore lowers the limiting thickness of scale; if the scale was porous the limiting thickness would be still further reduced. These calculations illustrate the desirability of preventing scale formation, particularly as the trend of modern boiler development is towards higher pressures.

Since there is a limiting temperature above which continued working under pressure will eventually lead to "creep," bulging and failure of the tubes, it is essential that conditions should be so chosen that the metal of the hardest worked tubes never exceeds this temperature. But the higher the pressure at which the boiler is operated the higher the temperature of the water inside, and so the smaller the margin that can be allowed for overheating. Moreover, it is just those places where the heat input is the greatest and the temperature drop through any scale is therefore a maximum, that tend to acquire a covering of scale. How serious such a problem may be is shown by a case recently described[2] in which failure of the furnace wall tubes of a boiler working at 1,300 lb. per sq. in. and evaporating 240,000 lb. of water per hour, occurred when they were covered with a scale only 1/32 in. thick. Where the economizer has been replaced by the air heater the resulting increased temperature of the fuel bed and furnace gases has led to the absorption of more radiant heat than hitherto by the boiler heating surface, with the same danger. The cooling of boiler brickwork and sidewalls by means of water tubes passing through them, and the water screens placed below the flame in pulverized fuel furnaces, mean an extension of the boiler heating surface, and their use (being in the hottest working zone and therefore exposed to severe radiation) compels the adoption of measures to prevent the formation of scale on such surfaces.

The only solution of the problem is to prevent as completely as possible the formation of scale. The ideal way would be to feed the boiler with pure water, and an approximation to this is to return the condensate to the boiler and employ distilled make-up. Even then, however, perfection is not attained, for the evaporator may prime and there is usually some little condenser leakage. From these sources sufficient scale-forming matter may be introduced to cause trouble in a high-pressure plant.

Most boiler deposits are composed essentially of compounds of calcium and magnesium. Calcium sulphate and carbonate, magnesium and calcium hydroxides and silicates are all found from time to time, but the most important components of boiler scales and sludges are usually calcium sulphate and carbonate. The different character of various scales has already been mentioned. Chemical analysis shows that the chief ingredients of hard, flint-like, firmly adherent scales are calcium sulphate and (more rarely) calcium and magnesium silicates ; the relatively soft, easily removable scales are generally composed of calcium carbonate and occasionally magnesium hydroxide.

Scale formation may also be divided into two groups—viz., that mainly due to the temporary hardness of the feed water, and that due mainly to the permanent hardness of the feed. When a solution of calcium or magnesium bicarbonate is boiled, bubbles of carbon dioxide are given off and chalk separates as a fine white powdery precipitate. If this be allowed to settle out and the water then decanted off it will be found that some of the hardness of the water has disappeared. This form of hardness is therefore called *temporary hardness*, because it is removed when the water is boiled. Hardness due to calcium and magnesium sulphate, on the other hand, cannot be removed by boiling, because the salts do not decompose and are quite soluble. Such hardness is therefore known as *permanent hardness*. In a boiler into which hard water is fed chalk will always be deposited by the removal of the temporary hardness, and if this chalk is allowed to remain it settles on plates and tubes and forms a scale. This temporary hardness may therefore be removed with the formation of a carbonate scale in feed water heaters, the water passing to the boilers with permanent hardness and carbon dioxide gas. The formation of scale from salts forming permanent hardness is not in general due to decomposition, but to the very slight solubility of the salts. Thus, 100 parts of boiler water will dissolve only 0·22 parts of calcium sulphate, and as soon as this small concentration is reached further amounts arriving with the feed simply deposit, usually in the form of a dense scale, on the hottest part of the heating surface.

Lime-Soda Process

The familiar lime-soda process of softening water was developed to deal with this situation. Clark suggested the removal of temporary hardness with lime-water as early as 1841, and the main difference in modern plants is that milk of lime is used instead of lime-water. This has an average concentration of 10 per cent as compared with 0·15 per cent for lime-water and therefore reduces the space required in plant for softening. The lime reacts with the bicarbonates, forming calcium carbonate and magnesium hydrate, both of which are insoluble and settle out in the settling tank, as shown by the following equations:

$Ca(HCO_3)_2$	+	$Ca(OH)_2$	=	$2\ CaCO_3$	+	$2\ H_2O$		
Calcium bicarbonate		Hydrated lime		Calcium carbonate (insoluble)		Water.		
$Mg(HCO_3)_2$	+	$2\ Ca(OH)_2$	=	$Mg(OH)_2$	+	$2\ CaCO_3$	+	$2\ H_2O$
Magnesium bicarbonate		Hydrated lime		Magnesium hydroxide		Calcium carbonate		Water
				(both insoluble deposits)				

The original method for dealing with the permanent hardness due to sulphates, chlorides, and nitrates of calcium and magnesium was developed by Dr. Porter, and consists in the addition of sodium carbonate to the water. This converts the soluble calcium salts into insoluble deposits, as follows:

$CaSO_4$	+	Na_2CO_3	=	$CaCO_3$	+	Na_2SO_4
Calcium sulphate		Sodium carbonate (soda ash)		Calcium carbonate (insoluble)		Sodium sulphate (soluble)

The only difference with other calcium salts is that a different sodium salt, e.g., sodium chloride, is left in the water. The permanent hardness salts of magnesium, such as the sulphates, can be removed by lime but, as shown by the equation below, the reaction produces an equivalent amount of calcium salt. This has to be removed by sodium carbonate, so that in effect sodium carbonate equivalent to the permanent hardness salts of both calcium and magnesium has to be used.

$MgSO_4$	+	$Ca(OH)_2$	=	$Mg(OH)_2$	+	$CaSO_4$
Magnesium sulphate		Hydrated lime		Magnesium hydroxide (insoluble)		Calcium sulphate (to be removed)

Other magnesium salts react similarly, the only difference being that different calcium salts are produced, e.g., the chloride, for subsequent removal with soda ash. In the lime-soda process as operated to-day these two chemicals are combined in suitable amounts to give the above reactions and so remove both temporary and permanent hardness.

If the theoretical quantity of chemicals necessary for softening be added to the water at atmospheric temperature it will be found that precipitation of calcium salts reaches its maximum only after long periods, and with magnesium salts substantial portions remain in solution. It is usual to add some excess of softening chemicals to make the softening reactions proceed more nearly to completion, and usually an efficiently operated cold-process lime-soda plant is capable of giving a softened water with residual hardness 2–4 deg. (parts per 100,000) with a modest excess of reagents.

Hot Lime-Soda Softening

The slight but measurable solubilities of calcium carbonate and magnesium hydroxide, and the formation of complex basic carbonates, limit the reduction in hardness achieved in cold-process softening unless special reagents are used, and even then a certain residual hardness remains in the treated water. The solubility of these minerals decreases as the temperature increases and, furthermore, rates of reaction rapidly increase at the same time. In addition, the viscosity of water decreases and is greatly reduced at temperatures of the order of 150°–200° F., which favours rapid settlement of precipitates. These are the chief reasons for the use of hot-process softening, and the conventional hot-process lime-soda softener operating with exhaust steam at up to 5 lb. gauge pressure and temperatures up to about 150° F. will give a residual hardness in the effluent of about 1 part per 100,000. Softening to this extent is adequate for most purposes, and if the works has a surplus of exhaust steam, efficient and very economical treatment can be secured in this way. A number of important practical advantages also follow. There are useful savings in chemical treatment. The high operating temperature results in thermal decomposition of bicarbonates to carbonate, and thus saves lime. It is found also that as the softening temperature is increased less excess sodium carbonate is required to obtain the same residual hardness, thus economizing in soda ash. This in turn means lower alkalinity in the feed, and together with the lower hardness in the treated water results in lower blow-down rates. The extent to which de-aeration is effected in a hot-process softener depends on the operating temperature, but it is usually substantial and leads to further useful savings of chemicals such as sulphite for oxygen absorption or reduces the work of vacuum de-aerators. The feed is also substantially preheated ready for any process of physical de-aeration, which saves fuel again. Further, a good deal of deposition in pipe lines, etc., from " after-precipitation " is due to the feed being heated after softening ; in hot-process softening this effect is minimized, and in addition the very small residual hardness offers much less potential deposit.

The operating temperature is limited to about 150° F. in conventional designs of plant owing to the convection currents set up by cooling at the walls of the softening tank. This results in faulty settlement of precipitates. To combat this factor, special features are incorporated in the plant, such as an inner hood or cone to protect an inner settlement zone from atmospheric cooling ; lagging of the exterior of the tank also is an obvious necessity in any fuel economy drive. Under such conditions, and with the softening tank closed to operate under pressure, the process has been successfully operated at 210°–230° F. In the past few years, however, pressure units operating at 15–20 lb. gauge (250°–259° F.) have been put into service, and even more recently units have been operated at 30 lb. (274° F.)[3] and 50 lb. (298° F.).[4] These two latter plants are quite large units, handling 70,000 and 28,000 gallons of raw water per hour respectively, and softening is carried out in two stages. In the first stage the water is given a treatment with lime and soda ash, and in a second compartment it is de-aerated and then treated with a new reagent, phosphoric acid, to reduce the hardness and alkalinity remaining after the lime-soda process. The final effluent is filtered, combined with condensate, and used as boiler feed without further treatment ; soap tests consistently show zero hardness.

A point to remember in connection with hot-process softening is the solvent action of alkaline waters on sand when heated; non-siliceous filters such as anthracite should therefore be used for handling the hot effluent.

General Notes on Plant

Available types of plant are numerous, although the basic principles of all are as outlined. The plants in use differ in the methods adopted to effect (1) preparation of reagents, (2) proportioning the reagents to the raw water, (3) mixing the reagents and the water, and (4) settling and filtering off the precipitates. The methods of accomplishing these stages may be considered somewhat generally and briefly, and in the first place it will be useful to point out the essential features of an ideal lime-soda softening plant. In such a plant the preparation of the reagents would be conveniently effected, probably automatically, and they would be made and used in such a manner that the concentration would remain unchanged until wholly consumed. The amount of reagent to be used in relation to the water would be accurately measured, and such accuracy would maintain for long periods of time. The measuring device should also maintain its accuracy at all rates of feed and yet be capable of easy and precise adjustment if required. Speedy and thorough mixing would undoubtedly take place in the ideal plant, and the capacity of the reaction tank would be such that ample time (probably at least 2 or 3 hours) would be allowed for completion of the softening reactions before the water passed to the settling tank. The filter would be efficient and readily cleaned, but in such an ideal process would actually be only supplementary to the efficient sedimentation already achieved in the settling tank.

(*a*) Preparation of Reagents.—Occasionally reagents are used in the form of a dry powder, purchased ready mixed. This is particularly suitable where a plant operates intermittently and a solution might evaporate or lose some solid by settlement in idle periods. In most cases they are used as solutions or suspensions in water, mixing or solution being effected in the smaller plants by hand and in the larger plants by mechanical stirrers or injector agitators, in a tank at ground level from which they are pumped to a supply tank. In many modern milk of lime plants it is customary to have both lime and soda ash in one solution. The lime required for treating a certain amount of water is dissolved with the requisite amount of soda ash and the whole diluted to a known volume. This solution of reagent is then allowed to pass to the reagent stock tank. Sometimes lime is used in the form of clear lime-water prepared in an automatic solutioner. This consists of a vertical tank, a large part of which is in the shape of an inverted cone. The feed pipe for lime passes down the tank almost to the bottom, and the tank is charged by passing milk of lime down this pipe. Water is then admitted at the apex of the cone and flows upwards. This results in agitation and solution of the lime, and any undissolved lumps and impurities are deposited as the solution passes up the cone owing to the gradually diminishing velocity as the cross sectional area increases. These accumulated impurities can be removed periodically by means of a sludge cock. The use of saturated lime-water obtained from the top of such a tower has certain disadvantages. The solutioner has to be of considerable capacity, and indeed will sometimes take about 25 per cent of the water to be treated, owing to the small solubility of lime. Again, sometimes the solution is not even saturated with this small amount owing to a coating of carbonate on the lime preventing solution. The softening process must also be carried out cold, owing to the reduced solubility of lime at higher temperatures. Finally, the proportion of lime-water used in the solutioner varies according to the temperature owing to the variation in the solubility of lime with temperature. It is not always convenient to be watching such an influence. On the other hand, it has the important advantage that additions of lime and soda are independent, and adjustments to the chemical dosing are therefore simplified. This is useful, particularly where a water is of variable composition. Sometimes a water varies a good deal throughout a single day even, and at such times it will often

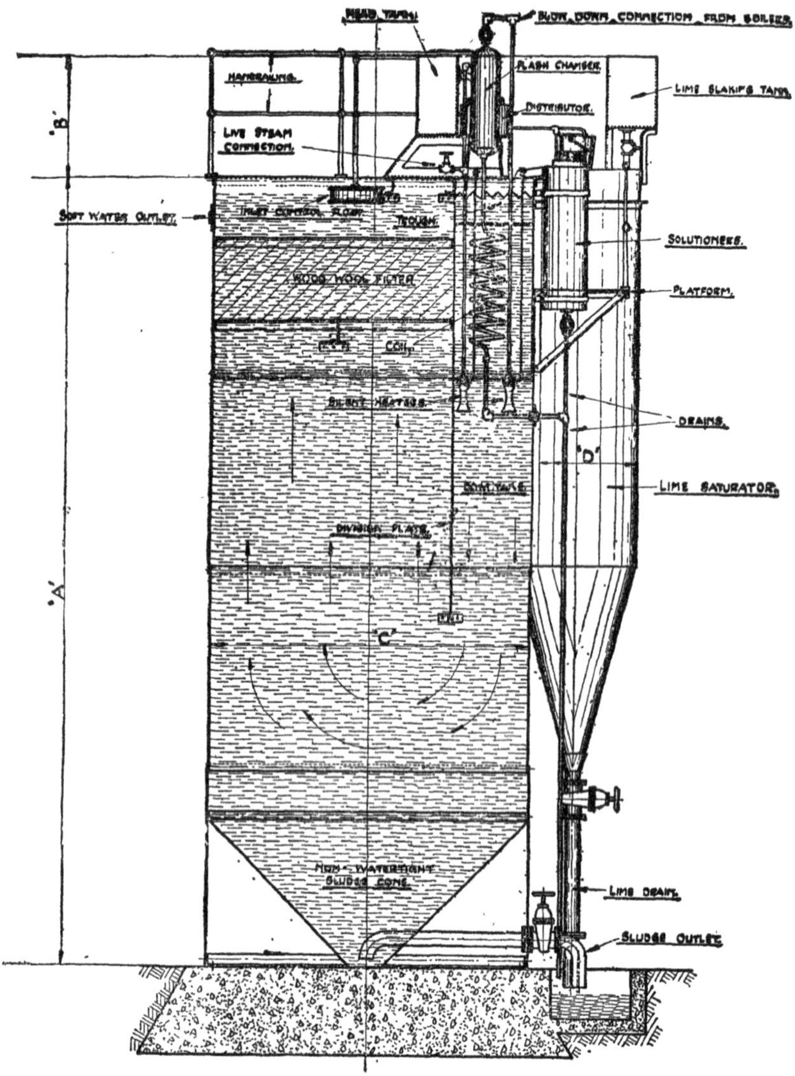

FIG. 1. Kennicott Hot Process Softener combined with continuous blow-down.

Hard water from the head tank is divided by a distributor into four portions: (1) Lime saturator, (2) Soda solutioner, (3) Alum solutioner, (4) Bulk of supply which passes to a mixing tray. The amounts of saturated lime-water or soda and alum solutions can be varied to suit any water by adjusting the distributor. A flash expansion chamber at the top of the softener can take continuous blow-down water from the boilers. Flash steam passes into the water in the softener via a vent pipe and a silent heater is also provided for raising the temperature of water as required.

be observed that the ratio of carbonate to non-carbonate hardness does not remain constant. This means that a different proportion of lime to soda ash is required for efficient softening of the water. With one feed tank for the lime and soda it is rather difficult to change the dosage of the chemicals quickly enough for such a situation. With separate lime and soda feeds each of the chemicals may be changed separately and more quickly. Of course, milk of lime plants may be designed with separate feeds for this purpose if desired.

(*b*) Measurement of Reagents.—With plants of the lime-water type the soda solution has to be made up and introduced separately. The previously prepared soda solution is then frequently displaced by the quiet inflow of a corresponding amount of water at the top of a "displacement tank," the soda solution running off through a pipe at the bottom of the tank. This then mixes with the lime-water and the water under treatment. In such a case the raw water is divided into one large stream and two smaller streams, of the proper size to make or displace the required quantities of lime or soda solutions. This division is effected by micrometer valves in the base of a common supply tank, by an adjustable dividing weir, or by a regulating cock on the inlet pipe by means of which any desired portion of the water may be diverted.

In plants using lime suspensions, with or without the addition of soda, the most usual arrangement for measurement is some form of tipping bucket operated by the main flow of water, and which actuates valves, scoops, or cups in such a way that a definite quantity of reagent is supplied to each charge of water which is passed. Some plants use double tipping buckets, and special devices are employed in most cases to increase accuracy (irregular or delayed tipping leads to irregular softening) or decrease wear and tear. Despite such devices, tipping buckets are often considered impracticable in large sizes, owing to heavy wear and tear. An exception is the firm of Messrs. William Boby, Ltd., who have constructed very large softening plants of proved efficiency, incorporating their special balanced tipper design. As a general rule, though, especially in the larger plants, makers prefer to subdivide the main flow of water so that only a portion passes the measurement outfit. The remainder of the water is frequently used to operate some form of stirring gear in the reaction tank. This allows several ways of adjusting the amount of softening reagents, viz., by altering the proportion of water passing to the measuring gear, by varying the concentration of the reagent, and by changing the volume of reagent added.

Some plants use a tipping bucket or water-wheel to deliver reagent in the form of dry powder, either from a hopper delivering to sliding drawers which empty themselves at each tip of the bucket, by continuous rise of a piston in a vertical cylinder which has a scraper to strike off the powder flush with the top, or by worm screw feed from the base of a hopper.

Dry chemical feeding may be adopted in place of the usual solution feed where large volumes of water require treatment. Cleaner and easier operation, with a neater and more compact lay-out, and avoidance of troubles such as evaporation of solution or settlement of solids in the idle periods of intermittently operated plant, are among the advantages claimed for this method of feeding. One design (see Plate 8) will dose water with any amount of chemical from 1 lb. per hour to almost 2 tons per hour. A revolving feed table rotates at constant slow speed below the stationary feed hopper, and a sectional feed knife deflects the requisite constant volume of dry reagent from the table into the adjoining vortex mixer. Arrangements are made for a wide adjustment in the rate of feed, and for preventing cavitation troubles. In the vortex mixer the chemical is immediately mixed with water and discharged to the required point of application. Conditions of varying water flow are met by proportionate variation in the rate of feed by some arrangement such as illustrated in Fig. 2.

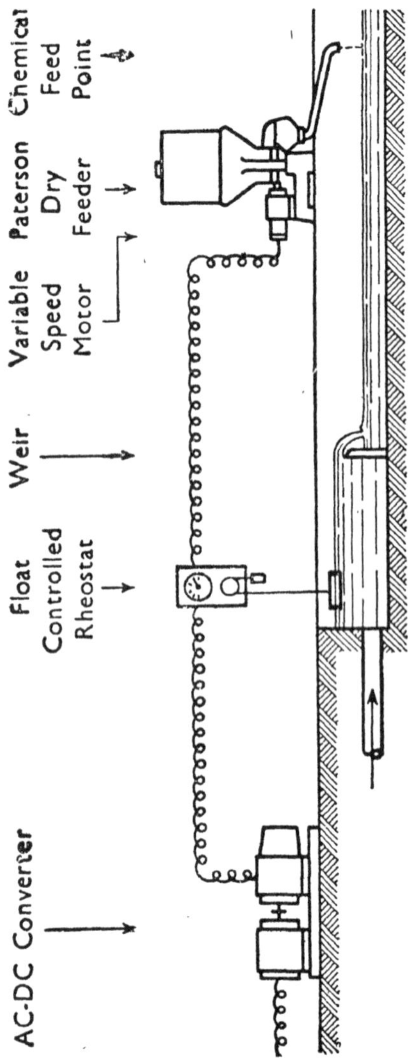

FIG. 2. Diagram of typical proportional Chemical Feed.

The most recently installed softening plants have the whole of the reagent storage and measuring equipment placed at ground level, and parts that require adjustment and control are no longer placed at the top of the settling tank. This greater accessibility has led to improved and more consistent results in operation. The development of a new electrical type of feed gear by Messrs. Permutit Co., Ltd., is a direct outcome of the demand for such ground control. In this gear, before entering the reaction tank the water passes through a water meter which contains a contacting device for operating a solenoid. When the meter has passed a prearranged quantity of water electrical contact is established and the solenoid operates a special valve, which releases a definite amount of chemical solution. The valve can be adjusted over a wide range for any discharge rate which, once fixed, remains constant irrespective of the head of solution in the reagent tank. This gives each measured quantity of raw water its correct amount of reagent irrespective of the rate of flow.

(*c*) MIXING AND PRECIPITATION.—Reagents and water usually flow together into the reaction tank, sometimes along a channel provided with baffles, and are thereby mixed. It is usual, too, to have some form of paddle actuated by a water-wheel. The water and reagent should always pass first to a reaction tank, where they are thoroughly mixed, and should then pass to the settling tank. Too often there is no distinction between settling and reaction tanks, and the reaction tank has to serve as a settling tank also.

(*d*) CLARIFICATION AND FILTRATION.—Separation of the precipitates should always be effected chiefly by settlement, during the passage of the water up or along the sedimentation tank, and completed by filters. If the water flows downwards, i.e., in the same direction as the solids, there will be poor settling. The rate of upward flow of water during sedimentation is very important, and usually the practice is not to exceed 6 ft. per hour. These tanks and filters determine the form and arrangement of the plant and are the basis of the classification of softening plants as "vertical" and "horizontal" respectively. The tanks should be so designed that the water will not be liable to short-circuit from the inlet to the outlet and so render a portion of the tank useless. This tendency is particularly prevalent with large tanks, which should be subdivided into compartments to avoid this by causing water to pass through the compartments in series or in parallel. With the horizontal type the actual filtering area is greater, but with the vertical there is more settling by gravity and therefore less work for the filters. The latter also occupy less floor space. Even after efficient sedimentation there will still be traces of precipitate to be removed, and the water is therefore passed through filters. Quartz sand filters have the advantage that they are more efficient and more easily cleaned than wood wool filters. Wood wool filters cannot be successfully used for down filtration, and in upward filtration it is advisable not to exceed 50 gal. per hour per sq. ft. of filter area. Downward filtration through quartz sand filters can be taken to 150 gal. per sq. ft. per hour. A point which should be borne in mind in connection with sand filters, however, is the possibility of dissolving some of the silica. This will certainly occur if the feed is appreciably alkaline, and it is advisable to keep the pH value below 10·0 in such a case. Otherwise silica dissolved in the water may react in the boiler with the residual hardness of the feed to form silicate scale unless methods of conditioning the water are adopted. Pressure sand filters are now quite common, cleaning being effected with an air-screw (compressed air at a few pounds pressure) agitating the sand and releasing deposits, which are then carried away by a reversal of the water flow.

Nuclear Precipitation

The most recent designs of plant embody radical changes in both mechanical and chemical features. Basically, these have resulted from the application of the principle of nuclear crystallization, which means that chemical precipitates form most rapidly and completely when there are present small crystals or nuclei on which the precipitates can grow. Such conditions result in formation of particles of much larger size than those obtained from spontaneous, undirected crystallization, and therefore settlement is much more rapid. In addition, with nuclear precipitation reactions can be brought to a definite stable end-point more rapidly, and this can eliminate such problems as the slow "after-precipitation" notorious as a cause of troublesome deposits in pumps, pipe lines, filters, and storage tanks in the older lime-soda plants. It follows, too, that these advantages enable complete softening to be effected in plants of considerably smaller size.

The value of contact with previously formed precipitate has been recognized for some years, and sludge return has been practised in some plants. The amount of nuclear material so provided, however, has been small and insufficient to obtain the full benefits of the principle. The technique is fully applied in the Precipitator (Permutit-Spaulding),[5] Accelator (Paterson),[6] and Spiractor (Permutit)[7] plants, in which either particles of previously precipitated sludge or sand grains are used as the nuclei. (See Fig. 3 and Plate 4.) The general method is to maintain a bed of the nuclear material in suspension by a combination of mechanical agitation and increased rate of flow. Precipitation takes place on the nuclei and as the particles increase in size they reach a point where they cannot be supported by the flow plus agitation provided, and therefore sink to a sludge concentration zone at the base of the plant for removal by automatic or other means. By proper control of the amount of nuclear material in use, the rate of flow, and extent of agitation to suit the water characteristics, a clear line of demarcation is formed between the upward-flowing treated water and the nuclear material, so that clear effluent containing only 5–10 p.p.m. suspended matter (even less under favourable conditions) is obtained. This can be removed in gravity or pressure sand filters, the low turbidity ensuring filter runs of considerable periods. It follows from these control conditions that plant performance is sensitive to load. To ensure that the plant always operates at the optimum design load, fluctuations in demand are met by cutting in and out on a high-low float control, so that the water flow is either zero or the optimum.

Despite the short time of contact allowed under the high flow rates required to maintain the nuclear material in suspension, it is found that the softening reactions are completed and a throughput of 2-3 gal. per sq. ft. per min. is obtained compared with the $\frac{1}{2}$ gal. for the normal old design lime-soda plant. While the conventional lime-soda plant requires

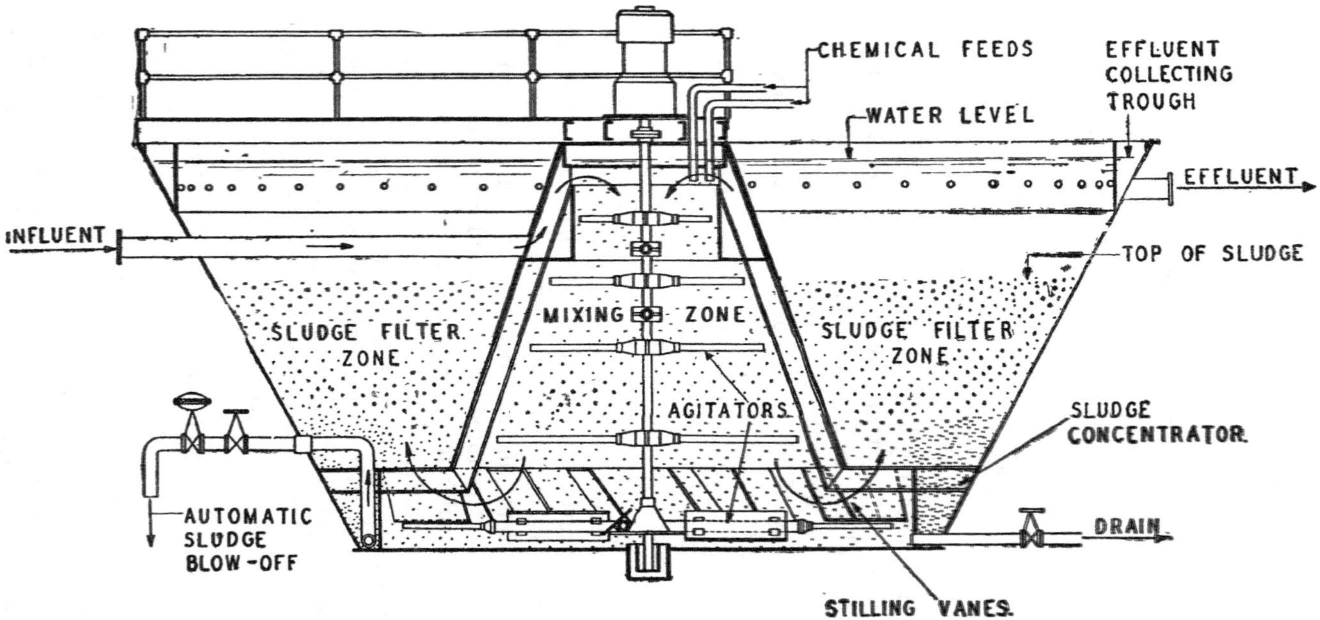

FIG. 3. Permutit "Precipitator"—sectional flow diagram.

a minimum period of 3–4 hours for settlement, ½–1 hour or less is sufficient with nuclear precipitation. Economy in chemical consumption is also achieved, since most of the dosage is actually used for softening, with the further result that lower excess alkalinity in the treated water curtails blow-down, etc.

Caustic Soda for Softening in place of Lime

Consideration of the reactions underlying the lime-soda process of softening will show that caustic soda can also be used instead of lime as follows:

$$\underset{\text{Calcium bicarbonate}}{Ca(HCO_3)_2} + \underset{\text{Caustic soda}}{2\,NaOH} = \underset{\text{Calcium carbonate (insoluble)}}{CaCO_3} + \underset{\text{Sodium carbonate}}{Na_2CO_3} + \underset{\text{Water}}{2\,H_2O}$$

This method of softening is in actual practice, chiefly in connection with the "sludge return" system. The advantages claimed for this system are that the condition of the water in the boiler itself governs the process, all heat from the blow-down is economically conserved, and purchase and handling of lime is avoided. The latter argument is a fallacy, since soda ash has to be supplied to give caustic soda by a process of decomposition in the boiler. In addition, by returning blow-down to the boiler the concentration of the soluble salts in the boiler will be unduly high and will eventually cause priming. Appreciable amounts of silica will find their way into the boiler owing to solvent effect of alkaline feed on the filters as outlined above. Finally, caustic soda is not a satisfactory reagent for the prevention of calcium sulphate scale, unless the concentration is high. Even if sufficient caustic soda is present to prevent deposition of calcium sulphate, there will be a very real risk of depositing calcium hydroxide (hydrated lime) as an adherent scale. The most suitable situation in which to use caustic soda as a softening agent in place of lime is where a natural water has a temporary hardness greater than its permanent hardness. When this is the case the whole of the caustic soda added is converted into sodium carbonate by interaction with the calcium and magnesium bicarbonate, as illustrated by the equation above. The use of the correct proportions of caustic soda will satisfactorily prevent scale in such a case.

Sodium Aluminate with Lime-Soda Treatment

The old lime-soda process of water softening, as originally applied, is subject to several objections. These objections include the residual hardness of 2–4 deg. which is found in the average effluent, the behaviour known as "after-precipitation," the incomplete removal of magnesium, etc. The use of sodium aluminate to overcome these objections was first proposed by P. W. Evans as early as 1922, when he showed that magnesium aluminate was readily formed as a flocculent precipitate with unique coagulating power. Recent research has focused attention upon this chemical, particularly with respect to its ability to remove silica as an insoluble calcium or magnesium aluminosilicate, and its adoption for water treatment purposes is rapidly increasing.

It will be convenient to trace the influence of sodium aluminate upon the various difficulties and failures of the lime-soda process by taking the problems in the order in which they are encountered in working operation of lime-soda plant.

The first source of trouble is the finely divided nature of the precipitates of calcium carbonate and magnesium hydroxide. The smaller the particles the slower will the precipitate settle, and consequently the longer the time required for efficient sedimentation. This means, of course, a larger settling space for a given throughput per day. It may be of interest to note that the average size of particles of calcium carbonate when precipitated at 210° F. is 21 times the size of those precipitated at 60° F., while the viscosity of water at 210° F. is only one-sixth of that at 60° F. The more rapid precipitation of hot-process softeners as compared with cold-process types is therefore understandable. The use of coagulants has sometimes been adopted to speed up sedimentation. These attempts were made with the sulphates of iron and aluminium, these salts combining with slaked lime to form flocculent precipitates of aluminium and ferrous hydroxide. These precipitates were found to coagulate the finely divided precipitate and also to remove any turbidity or colour which happened to be present. Unfortunately, these salts are acid and therefore corrosive, and this fact has diminished their use. It is argued that if the supply of lime and soda stops through any cause, the continuation of these acid coagulants would then have serious effects. Further, not only do these salts require lime to produce the floc (each lb. of acid coagulant requires about $\frac{1}{4}$ lb. lime), but they react with the lime to give calcium sulphate, the scale-forming salt:

$Al_2(SO_4)_3$	+	$3\ Ca(OH)_2$	=	$2\ Al(OH)_3$	+	$3\ CaSO_4$
Aluminium sulphate		Hydrated lime		Aluminium hydroxide (flocculent deposit)		Calcium sulphate (scale former)

This calcium sulphate has to be removed by addition of extra soda ash. The use of these coagulants therefore increases the amount of lime and soda required, and also increases the amount of soluble salts entering the boiler, apart from the corrosion risk outlined above. On the other hand, sodium aluminate forms flocculent precipitates with calcium and magnesium. These precipitates possess unique coagulating powers and, in fact, the floc so obtained produces better clarification than is obtained with aluminium and ferrous hydroxides. Furthermore, sodium aluminate is an alkaline salt; there is therefore no danger of corrosion—in fact, it acts as a softening agent by reacting with the calcium and magnesium salts. Its softening action is two-edged, as it were; it removes calcium and magnesium as aluminates and in so doing forms sodium carbonate (which, of course, is a softening agent) and thus reduces the soda requirements. It has been stated that a reduction in the lime required is also effected, the explanation being that the coagulant breaks up the colloidial interference of organic matter and double salts of magnesium, which so frequently retard the action of lime and prevent complete softening.

This leads to the question of the speed of the reaction, that is, the speed with which the precipitates are produced. The rate of reaction in cold water with normal lime-soda softening is notoriously slow, but with the new lime-soda-sodium aluminate softening the speed of chemical precipitation and also the completeness of softening reactions, both for hot- and cold-process plants, show a marked increase. The reduction in the hardness of the treated water is more rapid, and a lower degree of hardness obtained than with lime-soda alone. The following figures, expressed as parts of calcium carbonate per 100,000, illustrate the type of results which can be obtained:

Hardness	Alkalinity to Methyl Orange	Alkalinity to Phenolphthalein
1·7	4·6	3·1
0·7	6·0	4·5

In confirmation of these results, American investigators of the use of sodium aluminate in softening reactions state[8] that reactions are accelerated to such an extent that water containing less than 17 p.p.m. calcium and magnesium expressed as calcium carbonate may be secured at temperatures much below normal. Finally, data published by Spilman[9] may be mentioned. These show the uniformity of water treated by the lime-soda-sodium aluminate process, in spite of wide fluctuations in the raw water. An average reduction of 2 deg. in hardness and 1 deg. in alkalinity (both as grains per gallon) was obtained using sodium aluminate. Also, tests taken over a period of 3 hours indicated that the softening was practically completed within 1½ hours after agitation stopped, and clear water was produced within ¾ hour. This, of course, means a reduction in settling space, thus releasing room which can be more valuably filled by plant. It also lessens the risk of after-precipitation, because there is no delayed reaction between the softening agents and the hardness of the water.

There is another and more serious cause of "after-precipitation." This is the tendency of magnesium to form basic magnesium carbonates. These compounds are relatively soluble and may remain in the water for hours. They are, however, unstable, and may be precipitated on filter beds, pipe lines, etc. This "after-precipitation" shortens the length of filter runs by choking the filters more rapidly, reduces the internal diameter of pipes, necessitates frequent cleaning of economizers and heaters, and causes a great deal of trouble generally. Sodium aluminate reacts very completely and rapidly with magnesium to form a flocculent precipitate of magnesium aluminate. This compound is even less soluble than

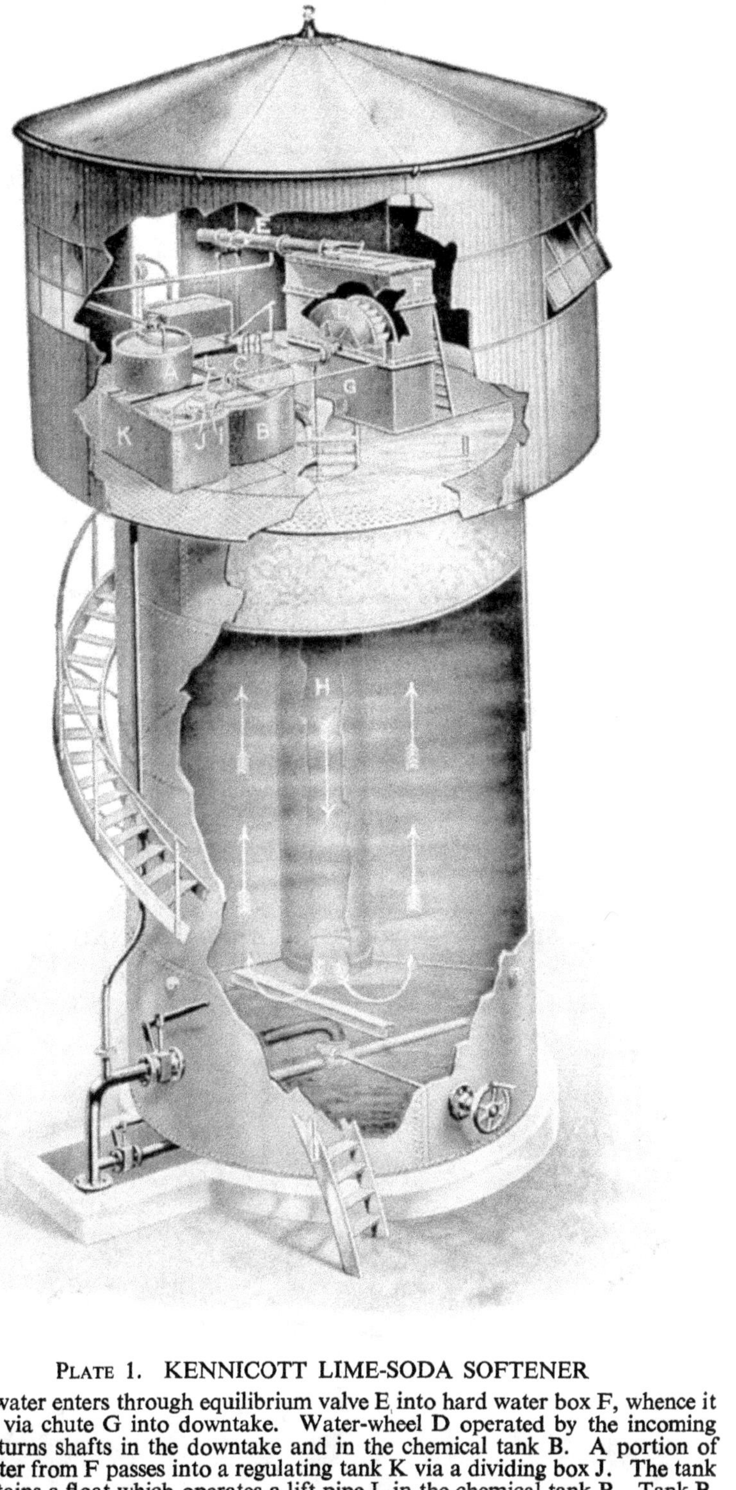

PLATE 1. KENNICOTT LIME-SODA SOFTENER

Hard water enters through equilibrium valve E into hard water box F, whence it passes via chute G into downtake. Water-wheel D operated by the incoming water turns shafts in the downtake and in the chemical tank B. A portion of the water from F passes into a regulating tank K via a dividing box J. The tank K contains a float which operates a lift pipe L in the chemical tank B. Tank B, by emptying in exactly the same time as K fills, thus supplies chemical in direct proportion to the water entering the plant. The water from the downtake moves upwards through the filter. Sludge removing gear is incorporated in the bottom of the tank.

PLATE 2. SECTION OF PRESSURE FILTER SHOWING UNDER-DRAINAGE SYSTEM

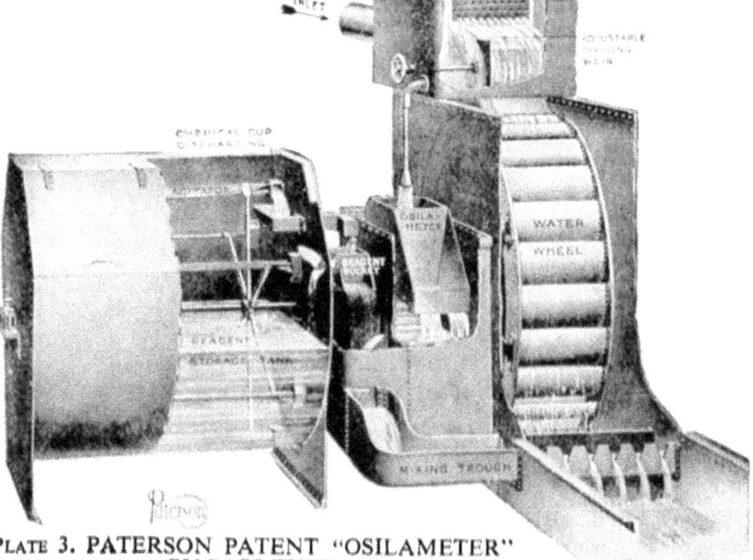

PLATE 3. PATERSON PATENT "OSILAMETER" (BY-PASS TYPE)

magnesium hydroxide, and treatment with aluminate results in a very low residual magnesium content in the treated water. This, therefore, avoids any tendency towards formation of basic magnesium carbonates and consequent after-precipitation of the magnesium. R. B. Beal and S. Stevens[10] have graphed figures for the cleaning period of filters in connection with a hot-process lime-soda softening plant before and after the use of sodium aluminate. Previously, the length of filter runs varied from 6 to 16½ hours, with an average of about 10 hours, but after sodium aluminate was adopted these increased to an average of slightly less than 20 hours. An actual case mentioned by these investigators is sufficiently interesting for quotation. One million gallons of water were passed through a cold-process softener in the hard-water district of South Yorkshire, and the water was softened from 20 deg. to 1 deg. (grains per gallon). After 3 hours' sedimentation it was passed through a quartz filter into a large concrete storage pond. Although stored in this way for three months no visible precipitation occurred, and samples taken every week showed no signs of precipitation when heated in a beaker to boiling point. Without the use of sodium aluminate the plant failed to soften below 3½ deg. without considerable over-treatment, and two days

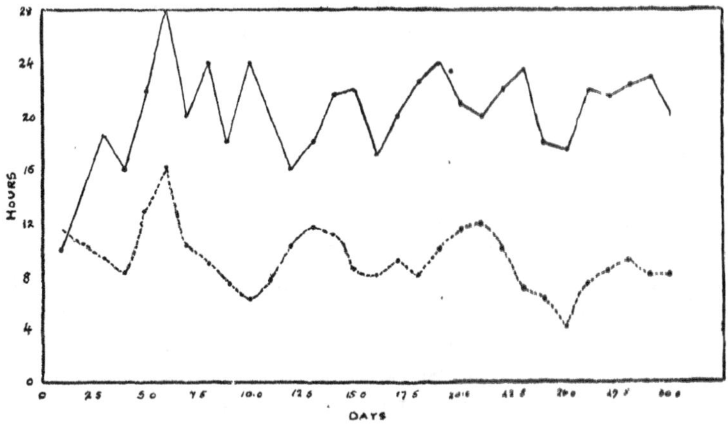

FIG. 4. Effect of Sodium Aluminate on time of filter runs.
Dotted line: without Aluminate. Continuous line: with Aluminate.

after the pond was filled the water became milky. The over-treatment which is frequently necessary with waters high in magnesia will, of course, result in a high alkalinity. This alkalinity will concentrate in the boiler very quickly and conditions favouring priming will soon be obtained. The use of sodium aluminate will lessen this danger by removing the magnesium salts without giving an unduly high alkalinity in the treated water.

The degree to which magnesium salts can be removed by sodium aluminate is shown in the following table of data[11] for the sulphate, chloride and bicarbonate:—

TABLE 3

Removal of Magnesium Hardness by Sodium Aluminate.

	Chloride		Sulphate		Bicarbonate	
	Zero	3 hrs.	Zero	3 hrs.	Zero	3 hrs.
Total alkalinity	—	1·1	—	1·2	—	1·1
Caustic alkalinity	—	0·55	—	0·55	—	0·85
Total hardness	—	2·9	—	3·0	—	3·0
Magnesium hardness	19·2	1·4	19·6	0·9	19·0	1·0
Calcium hardness	—	1·5	—	2·1	—	2·0

In the case of the chloride and sulphate the theoretical quantity of lime and soda ash was added, and then 3 parts per 100,000 of sodium aluminate. For the bicarbonate the theoretical quantity of lime was added before adding the aluminate.

The question of silica scale formation is quite as important as that of sulphate scale formation. Silicate scale is very hard and extremely difficult to remove, and is also a good heat insulator. The removal of silica is thus very important. Silica is found in water in both colloidal and soluble forms, most natural waters in this country containing $\frac{1}{4}$ to 1 grain per gallon. In conjunction with the usual hardness salts of calcium and magnesium it forms complex silicates; these form part of the silicate scale. With sodium aluminate it forms complex silicates which are insoluble in water and a valuable investigation of the reaction with aluminate was carried out some years ago.[8] This dealt with the reduction in silica in external softening by the use of sodium aluminate, and also with the effect in the boiler at high pressures. For the latter purpose an experimental boiler duplicating plant conditions was employed. The investigators found that if the water contained both magnesia and silica then sodium aluminate addition resulted in a marked reduction of silica with the formation of a flocculent precipitate of magnesium aluminosilicate. Sufficient aluminate must be added, over molecular ratio requirements, to leave in the treated water residual soluble alumina equal to the soluble silica.

Zeolite or Base Exchange Softeners

In certain geological districts there exist deposits of complicated silicates, such as sodium aluminosilicate, which are characterized by their property of exchanging the sodium radical for that of other metals when a solution of such a metal is passed through them. These complex silicates are known geologically as zeolites, or in water softening practice as " base exchangers " owing to their property of exchanging the metal bases of salts. This is the reason why some natural waters contain sodium carbonate, although the latter compound does not exist in a free state in nature. Silicates of soda and alumina are scattered through limestone formations and when a deep well water is found to be soft and highly alkaline it is due to the silicates or zeolites exchanging their sodium base for the calcium and magnesium bases dissolved by the water as it per-

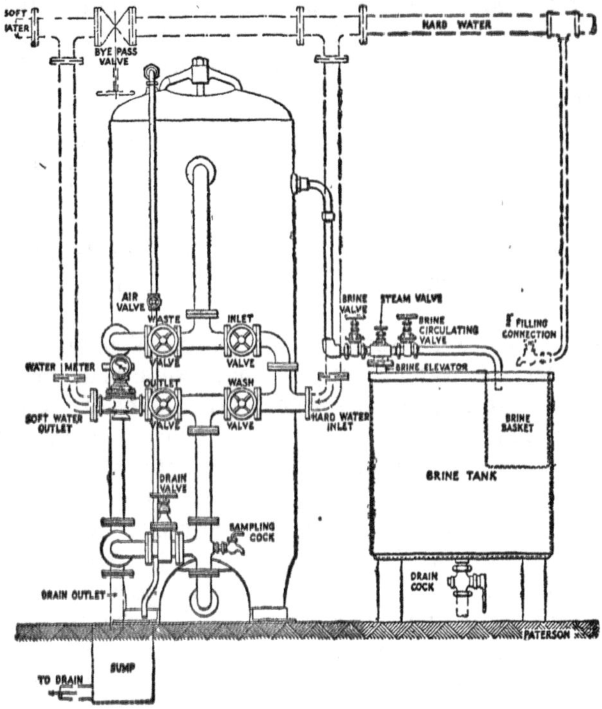

FIG. 5. Typical "Basex" Zeolite Water Softener Lay-out.

colates through the chalk and limestone strata. In such an action the calcium and magnesium elements are retained by the zeolite and an equivalent amount of sodium salts go into solution as follows:—

Na_2Z	+	$CaSO_4$	=	Na_2SO_4	+	CaZ
Active zeolite		Calcium sulphate (scale former)		Sodium sulphate (soluble)		Inactive zeolite where Z represents zeolite radical.

This reaction has been made the basis of a system of softening boiler feed water, as it will be appreciated that the hardness of water can be reduced to a very low figure. Originally the zeolites for such processes were those found occurring naturally, but they can now be produced synthetically in a greater degree of purity than the naturally occurring varieties. As would be anticipated, therefore, such artificial products have a greater exchange value per unit weight. This capacity or exchange value depends on the porosity or area of active surface in contact with the water. A porous zeolite requires more brine for regeneration owing to the absorption of the capillary passages.

It follows, of course, from the above reaction that after a certain amount of raw water has been passed through the zeolite for treatment it will become incapable of exchanging any more sodium for calcium and magnesium and the zeolite must then be regenerated before further softening can be effected. This is done by flushing the zeolite bed with sodium chloride solution, i.e., commercial brine—when the following reaction takes place:—

CaZ	+	2 NaCL	=	Na_2Z	+	$CaCl_2$
Inactive zeolite		Sodium chloride (brine)		Active zeolite		Calcium chloride.

The system must be cleared of calcium chloride formed in this reaction and a slight amount of water is wasted in effecting this clearance.

It may be shown by a simple calculation based upon atomic weights that although the soft water from a sodium zeolite plant may be reduced almost to zero hardness, the dissolved solid content of the treated water will be higher than that of the raw water. On an average there is an increase of about 5 per cent and with a raw water already high in saline matter this may be a serious consideration. On the other hand, in most cases water treated by the lime-soda method holds less solids in solution (by about 20-25 per cent) than before treatment, even when excess soda is used to complete reactions. This reduction is due largely, though, to the temporary hardness of the water, so that with a raw water low in temporary hardness this advantage of the lime-soda process is not so apparent. Even in this case, however, it should be remembered that the cost of removing temporary hardness by sodium zeolite is exactly the same as that for removing the permanent hardness, whereas by lime temporary hardness can be removed at something like one-fifth the cost of removing permanent hardness. Another point is that in the sodium zeolite process the bicarbonates are changed to sodium bicarbonate and then to sodium carbonate by decomposition by heat. This sodium carbonate accumulates in the boiler, and while it can be used for maintaining carbonate/sulphate ratios in the boiler, the tendency would probably be for the zeolite plant to supply too much sodium carbonate to the boilers and excessive blowing-down would be necessary to keep a reasonable concentration. With waters of high temporary hardness pre-treatment of the raw water would probably be imperative in order to remove the bicarbonates. This would be particularly true where the boilers operated at pressures above about 160 lb. since otherwise danger of caustic embrittlement of the boiler plates would arise owing to decomposition of the sodium carbonate. The high alkalinity produced by sodium zeolite softening of water high in bicarbonates can be reduced safely by means of sulphuric acid, with subsequent removal of the carbon dioxide by steam, and this method has been fairly widely adopted. An alternative is to treat the raw water with lime before passage to the zeolite. A final difficulty with zeolite softening is due to the fact that water of zero hardness has a great absorptive affinity for oxygen, such water taking up oxygen much more readily than raw water. Water from a zeolite plant should therefore never be allowed to cascade from the plant into a storage tank ; the outlet pipe should always be immersed well below the surface of the water. Zeolite plants are, however, very simple and easy to operate, no measuring being required, and where the raw water is of very good quality they are quite suitable. In other circumstances it is inadvisable to put a zeolite softener to remove the total hardness of the raw water ; the ideal is to use a zeolite softener to follow cold-process lime or lime-soda softeners.

Recent developments in zeolite plants have tended towards increased automaticity. It will be realized that in the control of a zeolite plant a number of valves are necessary to control the individual operations. Furthermore, the efficiency of the plant is impaired if backwashing, brining, and rinsing are not carried out at just the right time and to the right extent. In any case attention is necessary at such times. Messrs. Permutit Co., Ltd., have now introduced a completely automatic zeolite plant, illustrated at Plate 10. Essentially, this operates through a water meter, time switch, float switch, and a multiport valve operated by a motor. The first step in regeneration is initiated by the meter after a definite prearranged number of gallons of water have passed through the softener. Contacts in the meter actuate the motor driving the multiport valve, which then opens to the backwash position and the upward flow of water washes the bed until the time switch terminates this period. In the brine tank, meanwhile, the exact amount of saturated brine solution is automatically measured, and the right amount now passes through the softener owing to the multiport valve being opened to the suitable position after the time switch terminates the backwash. When this amount of brine has been fed, the level in the tank falls and a float switch in the tank shuts off the brine flow by causing the multiport valve motor to turn the valve to the rinsing position. A time switch controls this operation, after which the valve is again turned to the softening position. All these steps may be separately adjusted as required for individual cases, and no supervision is required. Several advantages of this type of plant will be realized. The softener never overruns its capacity, so that softening is always at full efficiency, and on the other hand waste of brine and rinse water by too early regenerations is avoided. Too early regeneration, of course, also cuts down the daily throughput of the plant. In addition slow turning of the automatic valve gives gradual increase in the rate of flow and thus minimizes disturbance of the zeolite bed. Finally, manually operated softeners are made large so as to avoid frequent regenerations, which consume the operator's time. With the automatic outfit the softener may be smaller and yet produce the same total daily throughput of soft water. This therefore effects savings in both first cost and operating cost.

Partial Zeolite Softening

It is sometimes found that operators of zeolite plants make a practice of passing some hard water around the softener with the idea of leaving the boiler feed with a little residual hardness, which will then form a thin " protective " scale as a counter to corrosion. The idea of a " protective " scale is an old one, and probably originates from the fact that many natural water supplies which are soft are also corrosive. The explanation for this is given elsewhere in this book (page 102) and it cannot be too strongly emphasized that if softening plant is properly operated a " protective " scale will be unnecessary. Intermittent feeding of soft and hard water also does not work well, since the soft water usually does not remove completely the scale deposited by the previous hard feed. Consequently scale gradually builds up to an undesirable extent. In addition, during the hard water feed periods corrosive elements will also enter the boiler, and at least intermittent corrosion will take place.

A controlled partial zeolite treatment supplies the boilers continuously with a properly adjusted blend of hard water and softened water, in such proportions that boiler deposits will consist mainly of sludge. The fundamental reaction which makes this possible is the fact that sodium zeolites react with the temporary or carbonate hardness of the water to form sodium carbonate or soda ash:—

$Ca(HCO_3)_2$ $Mg(HCO_3)_2$ Calcium and magnesium bicarbonates	+	Na_2Z Sodium zeolite	$\rightarrow$	$2Na(HCO_3)$ $2Na(HCO_3)$ Sodium bicarbonate	+	CaZ MgZ Calcium and zeolites

The sodium bicarbonate so produced will decompose when heated to give sodium carbonate (soda ash), and this in turn breaks down into caustic soda and carbon dioxide gas. These features of zeolite action are liable to be disadvantages, as discussed on pages 108, 110, but can be turned to advantage in this case by using the caustic soda and sodium carbonate as softening reagents. The soda ash will deal with the calcium sulphate as in a lime-soda plant; the caustic soda has reactions analogous to those of lime as shown for instance with calcium bicarbonate and magnesium sulphate:

$Ca(HCO_3)_2$ Calcium bicarbonate	+	$2\ NaOH$ Caustic soda	=	$CaCO_3$ Calcium carbonate (precipitate)	+	Na_2CO_3 Sodium carbonate	+	$2\ H_2O$ Water
$MgSO_4$ Magnesium sulphate	+	$2\ NaOH$ Caustic soda	=	$Mg(OH)_2$ Magnesium hydroxide (precipitate)	+	Na_2SO_4 Sodium sulphate		

Incidentally, it will be noted that the caustic soda reacts with temporary hardness to produce sodium carbonate, which of course can react with calcium sulphate, so that the caustic soda has a double effect.

Control is quite easy, since although the extent of caustic soda production depends upon boiler operating conditions, the whole operation is very simply regulated by hardness and alkalinity tests on the feed water. A feed water softening plant, properly operated, always leaves a slight excess alkalinity in the water to react with the residual hardness. Partial zeolite softening should be operated in the same way; the blending proportions should be selected so that the final feed contains a suitably regulated excess alkalinity over the hardness. Judgment of the amounts of such proportions is possible from the raw water analysis. The total alkalinity of the water includes all the temporary hardness plus any alkaline sodium carbonates present. The important point to bear in mind is that a zeolite treatment will not alter the amount of this total alkalinity, whereas the permanent hardness is eliminated. This means that a blend of soft and hard water will have a reduced total hardness, while the total alkalinity will remain unchanged. It is therefore quite simple to decide what proportions are necessary to ensure that the total hardness of the final mixture will be slightly less than the total alkalinity of the raw water supply. For example, suppose the raw water supply contains 18 g.p.g. hardness and 8 g.p.g. total alkalinity. Suppose, too, it is decided to operate with an excess of 2 g.p.g. alkalinity in the final feed. This will call for a dilution of 18/6, or 3 times, in the hardness. Mixture of 2,000 gallons of zeolite softened feed with 1,000 gallons of the raw hard water will then give a blend containing 6 g.p.g. total hardness and 8 g.p.g. total alkalinity. Since each grain of alkalinity is chemically equivalent to a grain of hardness, it is evident that the boiler is protected from scale deposition (and corrosion from minerals) with such a feed.

It should not be overlooked that the process is very largely an internal treatment, and therefore increases sludge formation in the boiler. This sludge may cause troubles such as localizing blistering of metal due to accumulation of sludge, foaming and carry-over, etc., and should therefore be removed by blow-down. A system of continuous blow-down is preferable, since sludge amounts are then more likely to be maintained at a suitable level, whereas with intermittent blowing-down periodical accumulation to undesirable amounts is inevitable.

There are a number of advantages to this partial softening procedure. Improved control of alkalinity reduces risk of embrittlement due to high concentration of caustic soda. Savings may be achieved in the size of softening plant required. Since only a part of the make-up has to be treated, a smaller unit will handle the daily throughput. In the case of plant to be laid down, this means saving in capital cost. On a softener already installed, running costs are reduced since regeneration is less frequent. Where plant is extending, and extra boilers are being laid down, the partial softening procedure may mean that the existing softening capacity will deal with the increased feed supply required without extra plant. Variations in the character of the raw water supply, of course, are countered by suitable adjustment of the proportions of hard and soft water.

Hydrogen and Demineralizing Zeolites

In recent years a great deal of research has been directed to the development of new types of synthetic zeolites. One of the objectives has been the production of a mineral with replaceable hydrogen, instead of the replaceable sodium as found in ordinary sodium zeolites. Minerals with replaceable hydrogen have formerly been produced, but unfortunately have been found unstable under acidic conditions. Similar instability is experienced with the sodium zeolites on exposure to acids, and as described elsewhere in this book continued contact of sodium zeolite minerals with acids or alkalis (even the slight amounts found in water supplies) leads to injury of the mineral and gradual loss of exchange power. The importance of this point is the fact that if a mineral with replaceable hydrogen is to be of any use, it will have to be regenerated with acid, since acids will be the only reagents capable of renewing the replaceable hydrogen. The new hydrogen zeolite mineral can withstand acids, and can therefore be regenerated with acids. The mineral is produced synthetically by the action of reagents such as fuming sulphuric acid, chlorsulphonic acid, etc., on carbonaceous materials. The origin of the names hydrogen or carbonaceous zeolites is therefore evident; the polymerized organic nature of the mineral enables it to withstand acid action, and therefore allows its use for the exchange of calcium, magnesium, etc., for hydrogen. The hydrogen zeolite reactions are as follows:

$Ca(HCO_3)_2$	+	H_2Z	→	CaZ	+	$2\ H_2CO_3$
Calcium bicarbonate		Hydrogen zeolite		Calcium zeolite		Carbonic acid (Carbon dioxide in water)
$CaSO_4$	+	H_2Z	→	CaZ	+	H_2SO_4
Calcium sulphate		Hydrogen zeolite		Calcium zeolite		Sulphuric acid
$CaCl_2$	+	H_2Z	→	CaZ	+	$2HCl$
Calcium chloride		Hydrogen zeolite		Calcium zeolite		Hydrochloric acid

Similar reactions are of course obtained with magnesium salts. The hydrogen zeolite is regenerated with acid as follows:

CaZ	+	H_2SO_4	→	H_2Z	+	$CaSO_4$
Calcium zeolite		Sulphuric acid		Hydrogen zeolite		Calcium sulphate

The mineral may be regenerated with hydrochloric or other acid also, or brine solution can be used; but if brine is used, of course, the mineral as regenerated will then contain replaceable sodium, not hydrogen.

Before the water softened in an acid zeolite cycle can be used for boiler feed, the free acids produced must be neutralized. This can be effected by addition of caustic or soda alkali to the effluent water, or by blending the treated water with more water softened with an ordinary sodium zeolite. The sodium zeolite effluent will contain a certain alkalinity from the temporary hardness, and this can be used to neutralize the acidity. Naturally, the carbon dioxide formed (as carbonic acid, or a solution of carbon dioxide in water) must also be removed if corrosion is to be prevented. According to conditions, amounts of the gas may be considerable. Some arrangement such as an aerating tower, open feed-water heater, or other de-aeration system will remove the bulk of the gas, with some supplementary treatment or de-aeration for dealing with the remainder of the gas and any dissolved oxygen if required. Usually the removal of dissolved carbon dioxide is effected quite simply by trickling the water down a tower packed with porcelain rings or wooden slats and up which a counter-current of air is passed. This stream of air strips the CO_2 content to about 0·5 part per 100,000, especially if

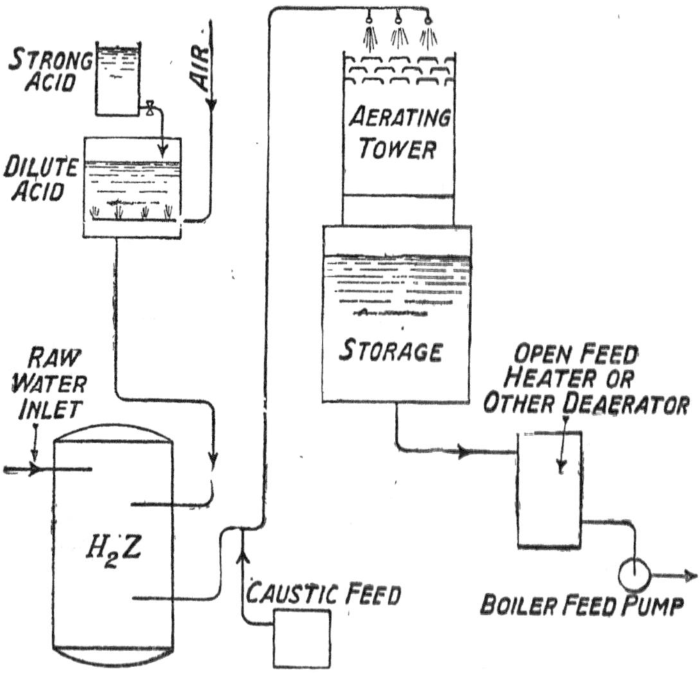

FIG. 6. Hydrogen Zeolite Plant Circuit with Alkali addition.

the water is first sprayed into the top of the tower. Recent plants have a cold vacuum de-aerator in place of the forced draught decarbonator. The advantage is that CO_2 is reduced to 2–3 p.p.m. and dissolved O_2 to 0·2 p.p.m., thus reducing corrosion of piping and need for resistant pipe materials. The type of plant circuit for alkali treatment and blending treatment is shown in Figs. 6 and 7.

By this means the carbonate content of high carbonate waters is reduced to an extent at least comparable with lime-soda softening. The method is specially valuable for the reduction of alkalinity found in some waters, since sodium carbonate and bicarbonate can be removed completely without leaving any permanent by-product. The reactions are similar to those for the corresponding calcium salts:

$$\underset{\text{Sodium Carbonate}}{Na_2CO_3} + \underset{\text{Hydrogen zeolite}}{H_2Z} \rightarrow \underset{\text{Sodium zeolite}}{NaZ} + \underset{\text{Carbonic acid}}{H_2CO_3}$$

$$\underset{\text{Sodium bicarbonate}}{2\,NaHCO_3} + \underset{\text{Hydrogen zeolite}}{2\,H_2Z} \rightarrow \underset{\text{Sodium zeolite}}{2\,NaZ} + \underset{\text{Carbonic acid}}{H_2CO_3}$$

Here again, of course, removal of the carbon dioxide is necessary before the water can be used for boiler feed. It will be evident from these reactions, though, that the alkalinity of a water can be reduced with the

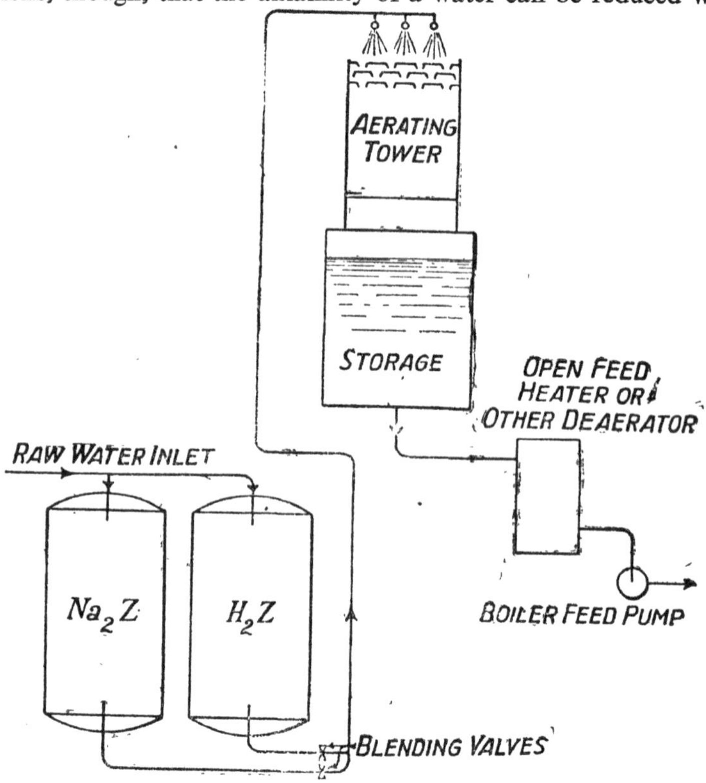

FIG. 7. Hydrogen Zeolite Plant Circuit when combined with Sodium Zeolite unit.

new mineral to any desired concentration. It may be noted, too, that the corrosion of condensate return systems, etc., is due to carbon dioxide; reduced contamination of steam with carbon dioxide should result from the above control of alkalinity. Under normal conditions, it is found that the acidity of the water from a hydrogen zeolite unit does not fluctuate but remains practically constant; this, of course, assists in the neutralizing process or the blending operation. Adjustable orifices in the effluent lines will maintain the proportions of H_2Z and NaZ waters constant, irrespective of rate of flow, and in the event of failure of the NaZ supply the H_2Z unit is automatically cut out to protect the system against acidity. Similarly, when flows are too low for accurate control, the plant is automatically shut down and feed is drawn from the storage tank.

It will be appreciated that although the sodium salts in a normal hard water will produce acids by interchange of sodium ions for hydrogen ions, the necessary treatment with alkali or blending with sodium-zeolite-softened water will correct this, and thus of course restore the water to its original state with respect to sodium salts. An appreciable reduction in the saline content results, however, thus minimizing blow-down, priming risks, etc., the extent of reduction depending on the amount of carbonate and bicarbonate in the raw water.

The next step, development of an acid-absorbent material capable of regeneration after repeated operation, originated in British research discoveries by Adams and Holmes.[12] These workers found that synthetic resins of the polyhydric phenol type would remove mineral acids, such as sulphuric and hydrochloric, while leaving the weak acids such as carbonic acid to be removed by degasification. Furthermore, the material could be regenerated with alkalies and used repeatedly. These early materials, however, had a comparatively low exchange capacity and a relatively short life owing to decomposition, etc. Intensive research since then has developed much superior materials, such as aliphatic amine resins, with very satisfactory life and high exchange capacity, the latter indeed being much higher than the capacities realized with the well-established sodium zeolites. In this connection, Myers and Eastes[13] have recently published some very interesting data comparing the capacities and efficiency of regeneration of natural zeolites, sodium zeolites, and the new synthetic resin materials.

It is evident that by combining the hydrogen zeolite process with a subsequent acid-absorbing process on the above lines, the effluent water will be contaminated only with dissolved carbon dioxide, which can be removed as already described. A completely demineralized water comparable with distilled water is thus obtained, as shown by the following table[14]:

TABLE 4

Average Demineralized Water Analysis compared with Commercial Distilled Water

	Expressed as p.p.m. Demineralized water	$CaCO_3$ Distilled water
Total hardness (Ca^{++} and Mg^{++}) ..	0–2	0–15
Sodium (Na^{+})	1–10	2–15
Total cations	1–12	2–30
Bicarbonates (HCO_3^{-})	1–6	2–18
Chlorides (Cl^{-})	0–3	0–6
Sulphates (SO_4^{--})	0–3	0–6
Total anions	1–12	2–30

Silica is not removed in the treatment, and supplementary treatment is necessary if its removal is essential. The cost of treatment to produce such demineralized water naturally varies according to the mineral content, but is far less than the cost of commercial distilled water.

Demineralizing plants have now been in successful operation on a wide variety of plants for several years,[15] the minerals having withstood some hundreds of cycles without appreciable drop in capacity (H-zeolite minerals are even more stable than the acid-removers). Normal practice is to regenerate hydrogen zeolites with 5 per cent sulphuric or hydrochloric acid and demineralizing zeolites with a 2 per cent solution of sodium carbonate. It should be observed that chlorides and sulphates are not entirely converted to their respective acids, and in fact it is found in practice that the acidity produced is about 90 per cent of theoretical. To increase this conversion, larger and uneconomical amounts of regenerating acid would have to be used in the regeneration stage. This is avoided by recycling a portion of the demineralized water to the inlet of the hydrogen zeolite unit, since it is found that the reduction in the theoretical mineral acidity achieved by this dilution materially affects the performance of the H-zeolite and results in appreciable improvement in effluent quality. Furthermore, normal acid consumption can be reduced by recycling.

Lime-Base Exchange Treatment

An excellent combination of the foregoing treatments is the use of lime to remove temporary hardness, followed by a base-exchange treatment to remove permanent hardness. This combined process eliminates the disadvantage of high alkalinity and carbon dioxide content in base-exchange effluent due to conversion of bicarbonates, and furthermore the lime treatment reduces the total solids of the effluent proportionately to the temporary hardness removed. Unfortunately, it is found that the cold lime reactions are completed too slowly, and the effect of filter sand grains or granular zeolite was to speed up the last stages of the reactions so that they were completed in the intermediate pressure filters or zeolite units. The cumulative deposition of these last small amounts of calcium carbonate led to gradual loss of filter or exchange capacity; in the case of the zeolite, the deposited carbonate could be removed only with an acid wash, which shortened the life of the mineral. The usual method of preventing this trouble has been to add small amounts of dilute sulphuric acid after the lime treatment, followed by passage through a marble chip filter bed to guard against excess acid entering the boilers.

One way of guarding against such delayed deposition is to use sodium metaphosphate to stabilize the lime-treated water so that the calcium carbonate is not thrown out of solution until the boilers are reached. Another proposal is to operate with organic zeolites instead of the sodium zeolite, since the former are resistant to acid and will therefore withstand the periodical acid wash required to remove accumulated carbonate. A still more recent development is the use of Spiractor plants for the lime treatment; owing to the new principles noted on p. 28, the reactions proceed to completion in this type of plant, and stabilized water is therefore obtained to pass on for base-exchange treatment.

The B.B.A. Dejector

In many works the absence of any orthodox pre-softening plant makes it necessary to protect the boilers by means of internal treatment, in which the chemical softening reactions are made to take place within the boilers, and when properly controlled very economical treatment can be obtained in this way. A small apparatus known as the B.B.A. Dejector is of particular interest in such cases, since the equipment also serves as a sludge extractor and prevents the undue accumulation of sludge which is usually a major criticism of internal treatments; in fact, in the usual internal treatment procedure there may be such excessive sludge production (and consequent danger of priming) that the addition of softening chemicals has to stop before all the scaling matter is converted to sludge.

The B.B.A. Dejector applies the principle of water circulation, the apparatus being connected to the boiler water space in a closed circuit in such a way that the boiler water circulates continuously through the apparatus as a result of thermosyphonic action. Piping connections are made so that this circulating water carries sludge from the boiler to the apparatus continuously. The sludge is thus steadily removed while it is still in suspension and is not allowed to settle on the plates, as occurs, for instance, in the intervals between intermittent blow-down. The amount of sludge that is carried into the apparatus by the circulating water naturally depends on the intensity of the circulation and on the sludge concentration in the boiler, and during years of research and experience the zones of sludge concentration in boilers and methods of obtaining the correct speed of water circulation have been specially studied to determine the piping arrangements required to give the best results under individual plant conditions.

The apparatus (see Figs. 8 and 9) is divided into two chambers; the upper chamber contains the chemical used in the treatment, while the lower chamber contains a central open-ended tube extending about half-way down the low chamber and surrounded by a special arrangement of concentric conical filtering baffles. A needle valve for regulating chemical dosage controls communication between the upper and lower chambers. The body of the apparatus is of mild steel, whilst the filtering baffles are of cast iron; the latter, incidentally, practically never need renewal or cleaning since any particles lodged on them are automatically blown off when the thick sludge is discharged from the base of the equipment. It will be observed that moving or delicate parts are eliminated, thus saving maintenance and repair costs. Owing to its small size, works space problems are avoided, and the equipment can be fitted easily to any boiler, whether stationary, marine, or locomotive, and can be removed without trouble from one boiler or boiler house to another.

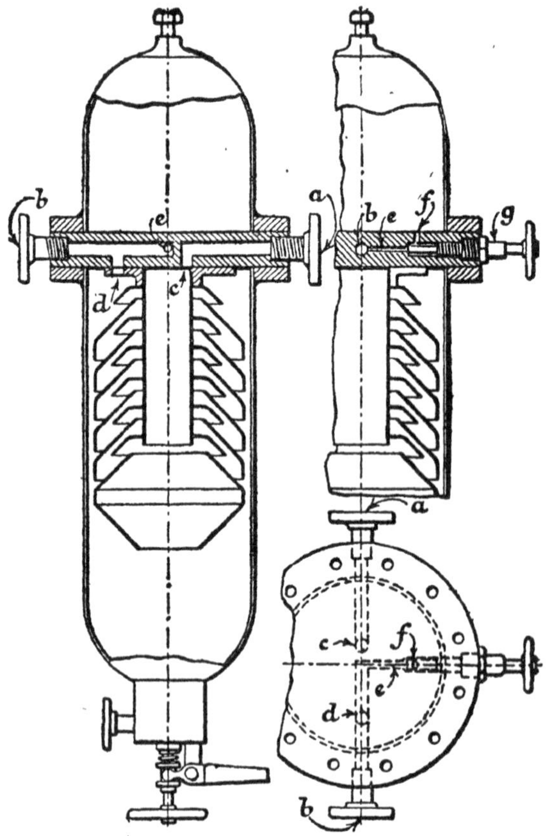

FIG. 8. Constructional details of B.B.A. "Dejector" water softener. (*Courtesy of "Engineering" and British Boiler Accessories Ltd., London.*)

Port *a* is connected to the boiler water space at its lowest point and *b* at a higher level. Port *a* opens into the central tube at *c*, and the clear water reaches port *b* via opening *d*. Leading into port *b* is a port *e*, which communicates with the reagent chamber through the opening *f*, the chemical dosage admitted being adjusted by needle valve *g*.

In operation the sludge-laden circulating water is made to pass down the central baffled tube into the return cone, where its direction of flow is reversed. At this point the heavy sludge particles drop out. The water then passes round the series of small and large baffles, where the remaining sludge particles are separated out owing to changes in direction and speed of flow. A slowly descending curtain of cooler and therefore heavier water prevents the main stream of the boiler water from rising through the annular space between the baffles and the shell of the apparatus. The sludge descends down the slopes of the baffles and along the wall of the apparatus to the bottom, where it accumulates and is discharged periodically as a thick mud by operating a lever valve. Before passing back to the boiler, the clear water receives a charge of softening chemical from the upper chamber of the apparatus; this dosage is admitted to the water as it leaves the separating chamber,

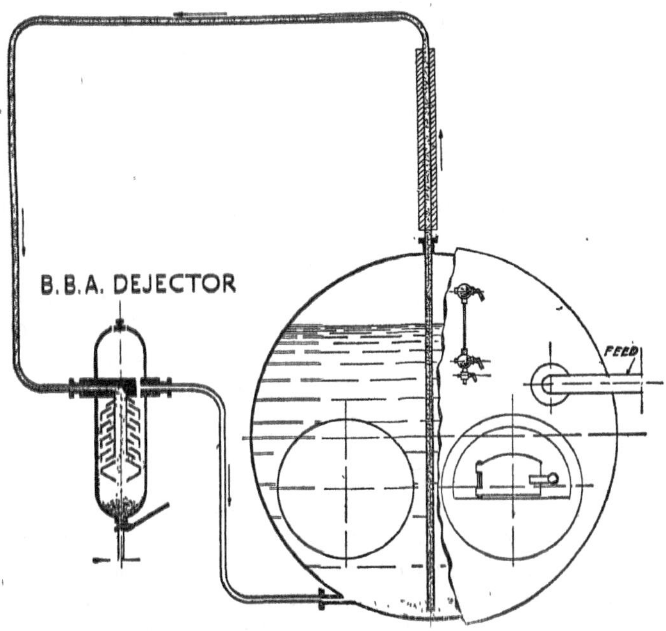

FIG. 9. B.B.A. Dejector—piping circuit.
(*Courtesy of British Boiler Accessories Ltd., London.*)

and is regulated by means of the needle valve, which is adjusted according to the alkalinity of the boiler water.

The raw feed water is admitted to the boiler in the normal way, the boiler heat breaking down the temporary hardness and forming a sludge of calcium carbonate. This sludge is steadily removed by the apparatus and the returning clarified water is dosed with soda ash to react with the permanent hardness inside the boiler, producing more carbonate sludge. If there is much magnesium in the water, caustic soda can be used as the treating chemical instead of soda ash; this alkali, of course, precipitates magnesia and in reaction with magnesium carbonates produces soda ash, which will soften any permanent hardness present. An economical double softening is thus obtained. Other chemical agents can, of course, be used if required ; for instance, a proportion of sodium bisulphite can be included to counter corrosion by dissolved oxygen. Again, if only certain boilers on the range require an additional chemical or special treatment, this can easily be arranged by introducing it only into those Dejector units serving the particular boilers in question (a battery of Dejectors in operation is shown at Plate 13). Where the plant includes an economizer, scale formation in the economizer can be avoided by adding a little tannin or sodium metaphosphate (see p. 64) to the feed water.

An important advantage of this method of treatment is that the alkalinity in the boiler water is readily controlled, since all that is required is regulation at the needle valve, with instant change in the chemical dosage passing to the boiler ; with external softening a temporary lag is likely, particularly if a large bulk of water has been treated and has to be fed to the boilers before the revised treatment can make itself felt. Furthermore, since the softening reactions are carried out under boiler conditions of pressure and high temperature, they are rapidly completed without excess alkalinity in the feed as is usually required with external softening plant. Economy in soda ash consumption is thus obtained, and in addition as already stated no chemical treatment of the temporary hardness is required, since the bicarbonates are thermally broken down under the boiler conditions into boiler sludge, which is then automatically removed and filtered out by the apparatus. Water tests are therefore simplified ; e.g., simply a daily test of alkalinity and a weekly test of hardness, readily done by unskilled labour, may be sufficient. The reduced chemical consumption and the improved control of alkalinity, coupled with the concentration of boiler sludge in the base of the apparatus, also lead to reduced boiler blowdowns, with accompanying fuel savings. In many instances it has been found that blowing down the Dejector has been sufficient to keep the concentration of soluble salts in the boiler water within safe limits. Finally, it should be noted that Dejectors have been operated successfully in conjunction with orthodox external softening plant, their role then being the removal of residual hardness left in the softened feed, and which of course passes to the boilers ; this aspect is of particular interest where plant changes have resulted in the external softening plant becoming too small and allowing a considerable amount of residual hardness to pass into the boilers.

Lime-Barium Softeners

Softeners using barium carbonate or barium hydroxide are sometimes recommended and are well known in America. In this process lime is fed to the water, after it passes through a bed of barium carbonate. This salt is not very soluble, but is sufficiently so to cause precipitation of all the sulphates present. Barium sulphate is produced, and this is

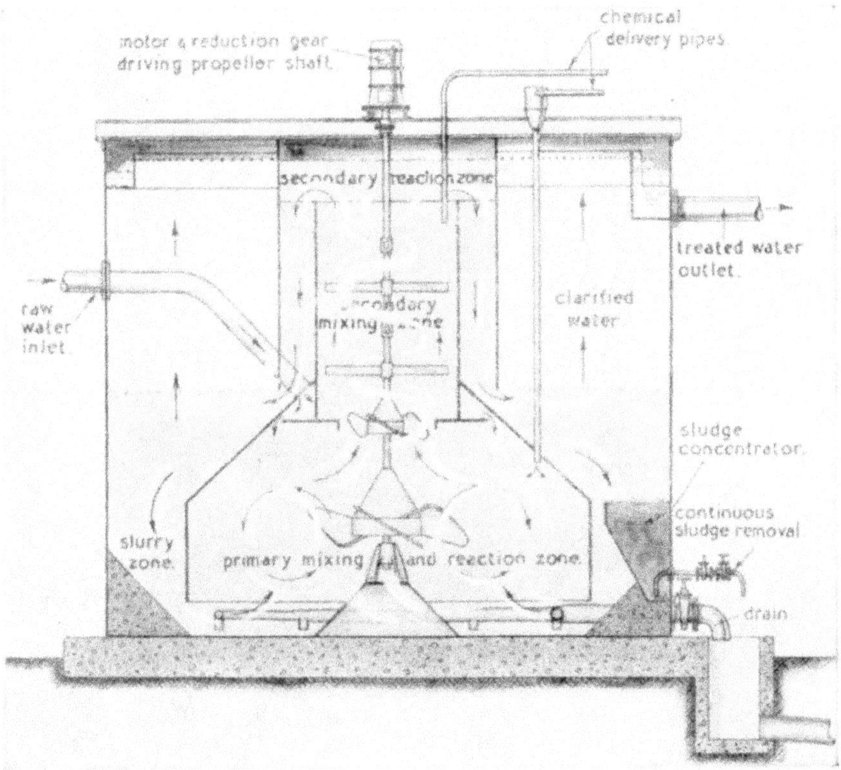

PLATE 4. PATERSON "ACCELATOR" FLOW DIAGRAM

PLATE 5. PERMUTIT "SPIRACTOR" PLANT
(Welton U.D.C. Capacity: 900,000 gallons per day)

one of the most insoluble substances known, so the removal of sulphate is very complete. The reactions taking place are as follows:

$CaSO_4$	+	$BaCO_3$	=	$BaSO_4$	+	$CaCO_3$
Calcium sulphate		Barium carbonate		Barium sulphate (deposit)		Calcium carbonate (deposit)

Na_2SO_4	+	$BaCO_3$	=	Na_2CO_3	+	$BaSO_4$
Sodium sulphate		Barium carbonate		Sodium carbonate		Barium sulphate (deposit)

The second reaction should be noticed. The replacement of sodium sulphate by the carbonate is not desirable, since it may lead to high caustic content of the boiler water and subsequent embrittlement of the plates and tubes. The double decomposition of calcium and magnesium sulphates by barium carbonate is accelerated by the presence of free carbon dioxide and the use of excess barium carbonate. The latter condition, of course, causes no technical difficulties since, being fairly insoluble, it can be used in the form of a bed as mentioned above and will remain there without loss of activity. It is important that the lime should be added after the barium carbonate so that the free carbon dioxide and carbonate hardness are not removed before the barium reactions. Barium carbonate does not, however, react with calcium chloride and this salt is left in the feed ; it may happen that the barium carbonate produces sufficient sodium carbonate by the second reaction above to deal with this, but the softening can generally be improved by secondary treatment with soda ash, which reacts with excess caustic lime and the chlorides, or better still, trisodium phosphate may be used with beneficial effects to be described later. The outstanding advantage of the barium carbonate process of softening is that the alkalinity and salt content are considerably less.

Barium hydroxide is sometimes substituted for the carbonate, in analogy with the use of caustic soda in the lime-soda process. It is not generally applicable, though, and does not remove chloride hardness. It has a wider use than the similar caustic soda process, which should only be used when the carbonate and non-carbonate hardnesses are equal. Barium hydroxide is best used under these conditions, but is also applicable if the carbonate hardness exceeds the non-carbonate hardness, since barium carbonate is only very slightly soluble ; if the relation is reversed, then barium hydroxide is unsuitable, since the treated water would contain caustic lime. The removal of sulphate with barium hydrate is more complete than with caustic soda, while the removal of carbonate hardness is about the same. Secondary softening with phosphate has about the same advantages with barium hydroxide as with the lime-barium carbonate process, but with the former loss of reagent in washing the purifier is avoided and a feed water containing the smallest salt content with the smallest residual hardness of all chemically treated waters is obtained.

Barium aluminate has also been used for feed water treatment and is suitable for alkaline waters. It prevents formation of silicate scale by absorption of silica in the same way as sodium aluminate.

Two great disadvantages of barium compounds for water treatment are their high cost and the fact that they are poisonous. Owing to the latter very careful handling is necessary.

Necessity for Conditioning of Boiler Feed

It will be appreciated from what has already been said that none of the foregoing methods of dealing with the boiler feed water problem of scale formation is a complete success. Lime-soda softeners leave a residual hardness of 2–4 deg. (parts per 100,000) in the water. Even with the adoption of sodium aluminate the hardness cannot be completely removed and is usually about one-third at least of the above figures. If the reactions taking place in zeolite softeners are considered it will be seen that, as the raw water contains sodium salts, some regeneration of the zeolite takes place continuously, and as a result the treated water will contain small amounts of calcium and magnesium salts. These will admittedly be small in quantity owing to the nature of the process, but such small quantities have proved dangerous under certain circumstances. Even when this regeneration effect of the zeolite is discounted, there is still the possibility of leakage, e.g., from the steam condensing system, of some raw water. This leakage renders treatment of condensate also advisable to counter the effect of the small amounts of scale-forming salts which are liable to pass to the treated feed.

Another point is the fact that a softened water having only two or three degrees (per 100,000) of residual carbonate hardness, *and no sulphate,* may yet form sulphate scale. This may seem extraordinary, but is due to the presence of sodium sulphate in the water. Recent research has shown that such a water may produce a scale due to the following reaction:

$CaCO_3$	+	Na_2SO_4	=	$CaSO_4$	+	Na_2CO_3
Dissolved calcium carbonate		Sodium sulphate		Calcium sulphate (scale)		Sodium carbonate

Further insight into this problem is given later when carbonate and phosphate conditioning and the ionic theory is discussed, but it will be immediately evident that in order to eliminate all possibility of the formation of a sulphate scale from a treated water it is advisable to give the water a further supplementary treatment. This is known nowadays as " conditioning " and several methods are available.

Colloidal Conditioning

One of the methods of conditioning consists in the addition of colloidal substances to the water and is known as colloidal conditioning. This form of protection has been known and practised for some time and the principles underlying its use are sometimes applied without any prior treatment of the raw water with lime and soda, etc.

It was shown by Graham in the latter half of the last century that there are two classes of substances which he called crystalloids and colloids. Crystalloids in solution will pass through a semi-permeable membrane such as parchment paper, while colloids will not so pass. This point is important in showing that a colloid presents a very large surface area or surface of contact. A colloidal solution may be described as having a substance in it whose constituent particles do not deposit or settle out, the particles being so finely divided that their mutual attraction is sufficient to overcome gravity. Such solutions may from all appearances differ little from true solutions in which the dissolved substance is resolved into molecular size. A colloidal solution of a given colloid is known as the sol of that colloid and when the colloid is coagulated it is known as a gel. Examples of gel formation are the solidification of the white of an egg upon boiling and the clotting of blood when treating a cut with alum. Glue, starch, tannin, gelatin, and casein are other examples of colloids.

The two important properties of colloids which are of service in the treatment of boiler water are adsorption and coagulation. These may be of service both in relation to scale formation and corrosion. With regard to scale formation, it is a fact that crystallization takes place most readily round a nucleus. Thus a supersaturated solution may be obtained by cooling a solution below its saturation temperature, but immediately a minute particle of solid matter is added, crystallization takes place very rapidly. The crystallization proceeds until the concentration of dissolved solids in solution reaches the normal saturation value. Minute crystals are formed in a boiler at first and these grow until gravity causes them to fall and adhere to the tubes and plates. The result is scale. When a colloid is present its adsorptive capacity is said to prevent this. Adsorption is a property which depends to some extent on the surface area and may be considered as a surface attraction (*ab*sorption takes place inside the absorbing substance). A finely divided substance like a colloid has a tremendous surface and therefore a high adsorptive capacity. The adsorption by the colloid of scale-forming salts deposited in the boiler is considered to prevent crystalline growth and therefore the formation of any appreciable amount of scale. It should be emphasized that this process is not chemical but purely and simply physical and may therefore be expected to operate equally well with any water. The other property of coagulation is also important in scale prevention. The coagulation of a colloid is an electrical phenomenon, since it is found that the majority of colloids carry an electrical charge when they are dispersed in sol form or form a colloidal solution. It has been stated here that the colloid is "dispersed" in sol form. In a colloidal solution there are two phases and the liquid medium is known as the dispersion medium, while the suspended particles form the disperse phase. It is the particles of the disperse phase which carry the electric charge, and if by any means this charge is neutralized, coagulation occurs and the colloid is precipitated in the form of a gel. This electrical precipitation is useful in a boiler, since neutralization of the charge may be effected by the scale-forming salts present in the water. In such a case mutual precipitation occurs, both the colloid and the scale-forming salts being precipitated as a gel *together*. The word together has been italicized because the gel forms a gelatinous film round the particles of scale-forming matter and thus prevents crystalline growth or scale formation. The spongy character of the precipitate is maintained since it contains such a large proportion of water that it will not bake on to the boiler plates or tubes when the boiler is working.

The exact mechanism of scale prevention may not be quite as outlined above, but undoubtedly any such prevention is due to the combined influence of the adsorptive and coagulant powers of the colloid. Both American and German research work confirmed this. Partridge, at Michigan University, showed that in the presence of tannin materials calcium sulphate crystals are much smaller and less regular. In short, in the presence of organic colloids crystallization was not as complete or rapid. In the German work the precipitation of calcium carbonate was studied in the presence of colloids such as dextrin and tannin. It was found that colloids in amounts from 0·1 per cent to 0·8 per cent substantially reduced the rate of deposition ; in one case the amount of precipitate was reduced by 70 per cent. The same effects were observed at pressures up to 150 lb., and colloid dosages as little as 0·1 per cent or less either totally prevented deposition or resulted in carbonate precipitation being in a non-adherent semi-colloidal form.

Colloidal conditioning is also said to reduce priming, the basis of the claim being the fact that suspended matter, whether impurities originally present in the feed or solids resulting from crystallization, is removed with the gel when the electrical precipitation mentioned above takes place. It will be shown later that priming and foaming are affected by the amount and condition of the suspended matter present and that the removal of such matter will reduce priming. With colloids in use there is therefore less tendency to priming and carry-over since the suspended matter is coagulated. The point has also been made that for scale prevention nothing like the quantity of chemicals theoretically required is used when colloids form part of the treatment. Normal chemical softening, for instance, will leave over a pound of sodium sulphate in the water for every pound of calcium sulphate removed. This adds to the density of the water and under evaporation the density may soon reach the point where priming obtains. Owing to the lesser quantity of colloid used, the density is not so high initially ; this is particularly true where no chemical pre-treatment is adopted and the addition of colloid comprises the entire boiler protection.

Because colloids have certain general properties it by no means follows that any colloid is suitable for use in a boiler. Hundreds of substances can be made colloidal, but all of them would not operate successfully in a boiler. Many colloids can be coagulated by the action of heat; these are unsuitable unless they require to be heated to a temperature higher than that to which they will be exposed in the boiler. It is essential that the colloid should be stable when the boiler is working and that any coagulation that occurs be due to neutralization of the electrical charges as outlined above. It should also be cheap and its condition should not present difficulties to commercial distribution. The starchy carbohydrates fulfil these conditions, but ordinary commercial starch is not suitable, since it does not give a truly colloidal suspension. Nowadays various oil-bearing seeds, such as linseed and cotton-seed, are used, the starch being extracted in specially constructed apparatus. The question may be asked why oil-bearing seeds should be used when oil is known to have a bad effect in boilers. In the Filtrator apparatus, for instance, which uses linseed, this difficulty is avoided by the fact that the method of extraction does not permit the extraction of oil. This is due to the fact that the oil is in the centre of the seed and will not be extracted if the seeds are unbroken. In the Filtrator system there is no crushing action upon the seeds and they remain unbroken. Experimental work has been carried out in this connection and a figure of the order of 0·03 per cent has been published as the percentage of total oil content so extracted. This amount is negligible and, indeed, it has been suggested that it may have the same effect as castor oil in reducing priming. It is also a fact that colloids often serve to maintain the oil in emulsion form and prevent the aggregation of the finely dispersed globules to a dangerous size. Finally, in the use of any colloid it is advantageous that the method of feeding should take account of variations in the rate of supply of the feed water. Variations in steam demand would then be met with increased supplies of extract to the feed.

So many conditions influence the action of colloids that they are apt to fail most disconcertingly, and so little is known about the effect of various conditions that one cannot always rely on a given colloid in a given case, although the same colloid may have functioned admirably in an apparently similar case previously. A more usual procedure is therefore to use a combination of colloid and inorganic chemical for internal treatment.

One great advantage of this combination is that it seems to be unnecessary to use the theoretical quantities of the alkaline salts, thereby lessening the cost of chemical treatment. This is due to the reaction of the alkalis and inorganic reagents in the mixture with the temporary hardness of the water. The ordinary softening reactions occur, and it will be realized that in addition to removing temporary hardness these reactions produce sodium carbonate, which can then react with some of the permanent hardness in the water. A double or two-edged softening is thus accomplished economically. The action of the mixture can probably be summarized as follows. Partial softening will first occur due to the above normal softening reactions, and the precipitates thrown out of solution will probably be partly coated with colloid which will prevent their forming adherent deposits on the heating surfaces. Any residual hardness in the water will be present in an unsaturated solution, and therefore scale-forming matter will remain in solution. When the solution becomes saturated and in a condition to deposit scale, the colloid will exert an adsorptive effect on the scale-forming matter as it comes out of solution. Scaling should therefore be prevented. A complex precipitate of colloid-scale salt will thus be deposited at all stages in evaporation and this, as already mentioned, will not adhere to the boiler metal but remain as sludge in the water. The colloid and sodium salts may be added together continuously to the feed water, or at frequent intervals, either in the solid state or dissolved in hot water. More accurate control can be obtained in the latter case. One precaution *must* be observed. The colloid should be used in amount sufficient to leave a slight excess in the boiler water. This is vital. The colloid has no injurious effect on the metal and a deficiency of it can mar the whole process.

It was mentioned above that with such treatment less than the theoretical quantities of alkali salts could be used. The actual proportion of the theoretical amount which will be necessary will vary fairly widely from one plant to another, according to feed water composition, rate of evaporation, type of boiler, etc. In some cases a quarter of the amount of softening salts calculated for a particular feed supply will suffice, but the optimum treatment can be determined only by trial. The slight excess of colloid which is imperative will not be nearly as expensive as the softening salts which would be required in the absence of the colloid.

Some judgment of the relative proportions of the alkaline and inorganic reagents, and the amount of such reagents to include in the mixture, as well as the dosage for the boiler, may be made by considering the ordinary softening reactions. The facts to bear in mind are: (*a*) Caustic soda reacts with temporary hardness to produce calcium and magnesium precipitates and sodium bicarbonate, the latter producing sodium carbonate; (*b*) the sodium carbonate so formed, and also that present in the mixture, can react with the permanent hardness to form precipitates and soluble sodium salts; (*c*) caustic soda will also react with magnesium salts to produce magnesium precipitates and soluble sodium salts. As a result of this there are three possible methods of procedure:

(i) Where the temporary hardness is such that its reaction with caustic soda would yield just sufficient sodium carbonate to react with all the permanent hardness. In this case a slight excess of caustic soda over that required for such reactions can be incorporated with the scale-preventing medium to give the boiler water an alkalinity.

(ii) Where the temporary hardness is such that its reaction with caustic soda would yield more than enough sodium carbonate to deal with the whole of the permanent hardness. It will be sufficient here to have enough caustic soda in the scale-preventing medium to react with such temporary hardness as will produce sodium carbonate to counteract the permanent hardness. The excess temporary hardness can be dealt with by the scale-preventing medium as it becomes thrown out of solution as evaporation proceeds.

(iii) Where the permanent hardness is in excess and cannot be handled by sodium carbonate yielded by the whole of the temporary hardness. Sufficient caustic soda is required to react with the whole of the temporary hardness with a slight excess for alkalinity—the sodium carbonate from this may be considered to react with part of the calcium sulphate, the remainder being taken care of by the scale-preventing medium. If the permanent hardness includes magnesium sulphate and chloride it must be remembered that these salts will react with both sodium carbonate and caustic soda and allowance must accordingly be made for this, since these salts may give rise to corrosive acids.

If other inorganic reagents such as phosphates are included in the mixture, their reactions with ordinary hardness salts should also be taken into account.

Individual conditions may demand some modification of these methods, but if treatments are modified on the above lines based on water analysis appreciable economies in chemical treatment may be effected. This has the advantage also of reducing the amount of dissolved solids in the boiler in addition to assisting control of the alkalinity. It is assumed, of course, that the colloid scale preventive works satisfactorily.

Tests should be carried out weekly with standard acid and soap solution and if the figures obtained are correlated with the state of the boiler as revealed by periodical examination the most efficient and economical treatment can be evolved. If the alkalinity has increased one week a little less treatment can be given the next week, since the increase in alkalinity means that the boiler has been getting excess treatment and if this continues the alkalinity will eventually reach such a point that free caustic soda will appear in the boiler. There is then danger of priming and caustic embrittlement. A good point at which to keep the average low-pressure boiler would be about 6 to 8 deg. hardness. If the hardness has risen one week owing to slight undertreatment there may be some slight deposit formed on the heating surface. A slight excess of treatment the next week will take care of this, since existing scale tends to disintegrate when the feed water no longer adds to it, and this disintegration process is hastened when colloids are present.

Various colloids are available for use with alkaline salts. There are the tannins, cambier, the galls, campêche wood, quebracho, the cutches, and celluloses such as dextrin. One cannot state exactly how much colloid to use—here again trial is the best way to find out. As a guide to general practice it may be recommended that ½ lb. of colloid be used for every 2 lb. of softening salts, though while one feed supply may need less than this another may take 1 lb. of colloid. On the French State Railways a 1 : 1 mixture of campêche wood and soda ash is thoroughly boiled in water and the solution added to the feed.

A disadvantage of internal treatment by these means which must not be overlooked is the accumulation of sludge in the boiler. If this is tackled by systematic blow-down, however, no trouble and very little inconvenience will result. The time of steaming necessary to bring the boiler to a maximum concentration of total solids suitable for efficient operation of the unit can be readily calculated, thus allowing the intervals between blowing-down, and also the amounts of blow-down, to be specified to the boiler operator. This will avoid undue concentration of the boiler water and also undue waste of hot water.

Interesting examples of colloidal treatment in industrial boilers have been described.[16] Treatment was with a paste mixture of about 60 per cent caustic soda and 40 per cent disodium phosphate concentrated in tannin extract until the final mixture contained 30–35 per cent extract. Dosage was ¼–½ lb. mixture per 1,000 gal., either as a supplement or a substitute for normal external softening ; this " starvation " type of treatment in which the alkaline dosage is of the order of a quarter or a third of that theoretically needed, has been regularly and continuously used with tannins particularly. On the above plants, boiler troubles due to incomplete softening at the lime-soda plant (caused by sudden variations in a river supply) were eliminated and excellent alkalinity conditions maintained, while old deposition in boilers and pipe lines was gradually removed. A marked reduction in priming was also realized, accompanied by pronounced improvement in lubrication and reduced cost of same.

The marked effect of tannins on old scale deposition is very interesting. The action is perhaps best likened to a rotting action that weakens the scale and accelerates its disintegration. Whether this is due to some formation of unstable compounds between the tannins and the calcium and magnesium is not known, but whatever the destructive mechanism there is little doubt that if a tannin concentration of appreciable amount —say about 8 g.p.g.—is carried consistently in an alkaline boiler water the character of existing scale deposition often changes gradually to a soft, dark-coloured, and easily removed non-crystalline deposit.

It will not be out of place here to note that the field of organic chemistry generally has been almost ignored in connection with boiler feed water treatment, the only organic preparations used in boilers generally being classed as " boiler compounds," which are not always regarded favourably by power plant engineers. As a result, organic matter as a whole is condemned. The use of properly compounded material has been successful in many cases, and from the nature of these it is evident that there are considerable possibilities in this field. Apart from any colloidal nature, these possibilities are largely due to the susceptibility of organic matter to pressure and temperature; under boiler pressure and temperature complex decompositions and reactions take place with beneficial results to boiler operation. This is more evident in connection with corrosion and caustic embrittlement inhibition, though the physical properties of colloids are of value in scale prevention.

Mechanism of Scale Formation

A much-needed investigation of the mechanism of scale formation was carried out some years ago jointly by an American manufacturer of boiler plant auxiliary equipment and the U.S. Bureau of Mines.[17–22]

As a result a theory of scale formation was developed on the following lines:

When water containing dissolved salts is heated, the solution becomes more concentrated as the water passes off as vapour, and eventually the dissolved substances will pass out of solution. There are two classes of substances, (i) those whose solubility in water increases with increasing temperature, and (ii) those whose solubility decreases with increasing temperature. It will be appreciated that the result of heating a solution of either of these salt types is the precipitation of the substances in solution. R. E. Hall's theory was that all salts, the solubility of which diminishes with increasing temperature, will form a scale on heating surfaces. Some experiments by Partridge and White, of the University of Michigan, may be briefly described to show the basis of this idea. They examined with a photomicograph outfit the deposition of scale on a polished chromium heater in a boiler vessel containing solutions of various amounts of calcium sulphate. They found that the initial deposition of scale occurred when a bubble of steam or air was formed on the heating surface.

Fig. 10 shows such a bubble formed on the heating surface. The bubble will temporarily insulate a part of the plate from the bulk of the water, and that part of the plate will therefore become overheated. Under the edges of the bubble (at x and y) there will be only thin layers of water between the plate and the bubble, and these thin layers will also become overheated. If the water was initially saturated with calcium

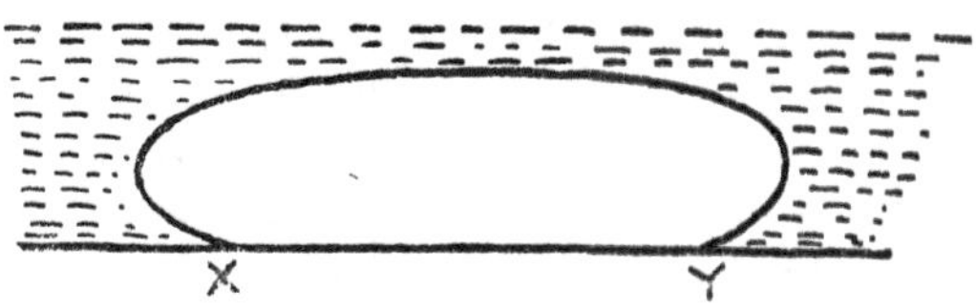

FIG. 10. Scale Formation at Steam Bubble.

sulphate then at x and y it will become supersaturated with this substance, since its solubility decreases with increasing temperature. The result will be the deposition of a ring of crystalline calcium sulphate around the edge of the bubble. After the bubble has burst, more calcium sulphate will be deposited on this ring of crystals for the following reasons: (*a*) the contact of the boiler water with the overheated portion of boiler plate will cause supersaturation; (*b*) as evaporation proceeds crystal growth will take place.

It was therefore said that substances such as calcium carbonate do not form scale because they are more soluble as the temperature increases; that the solutions at *x* and *y* around the edge of a bubble would be unsaturated (due to the overheating) with respect to such substances as calcium carbonate; in short, that such substances would deposit from solution in the cooler body of the boiler water, and thus form sludge rather than scale.

This seemed to give a reasonable explanation of why some substances form scales and others form sludges, but unfortunately research[23] disclosed the fact that calcium carbonate, too, diminished in solubility as the temperature increased. This discovery, coupled with the fact that carbonate scales found in practice are usually thin, soft, and easily removable, has led to various modifications and suggestions in the mechanism of scale formation. Differing crystal forms which vary in their ability to knit together and form scale; differing solubilities which involve differing amounts of a substance being in the area of the heating surface for scale formation; high rates of evaporation resulting in increased scaling due to steam bubbles being formed too continuously to allow crystal deposits to be washed away after release of each bubble of steam; differences in the amount of solubility diminution with temperature rise; these are some of the factors which have been suggested as perhaps influencing the different scaling behaviour of calcium sulphate, calcium carbonate, and other salts.

Irrespective of the explanation, however, it is a fact that in practice calcium carbonate precipitation is preferable to that of calcium sulphate. It is to ensure that this will take place that what are known as carbonate and phosphate conditioning processes have been developed.

Carbonate and Phosphate Conditioning

The term "boiler water conditioning," referring particularly to phosphate and carbonate conditioning, simply means that the condition of the boiler water is arranged so that the less troublesome impurities (such as calcium carbonate and calcium phosphate) are precipitated before or in preference to the more troublesome impurities (such as calcium sulphate and silicate). A brief consideration of solubility on the basis of the ionic theory will show how this is achieved.

According to the ionic theory, when a salt is dissolved in water it becomes partially dissociated into its constituent ions and the solution will contain oppositely electrically charged ions and also some non-ionized or undissociated salt. If the solution is saturated and in contact with some undissolved salt, equilibrium will be set up between these and the undissolved salt. This state of affairs may be represented graphically as:

$$\begin{array}{lcccccccc} & AB & \rightleftarrows & AB & \rightleftarrows & A+ & + & B- \\ \text{or} & A_xB_y & \rightleftarrows & A_xB_y & \rightleftarrows & xA+ & + & yB- \\ & \left(\begin{array}{c}\text{undissolved}\\ \text{solid}\end{array}\right) & & \left(\begin{array}{c}\text{dissolved}\\ \text{solid}\end{array}\right) & & \left(\begin{array}{c}\text{positive ion}\\ \text{or cation}\end{array}\right) & & \left(\begin{array}{c}\text{negative ion}\\ \text{or anion}\end{array}\right) \end{array}$$

In the case of sodium sulphate the conditions would be represented as:

$$Na_2SO_4 \rightleftarrows [Na_2SO_4] \rightleftarrows 2[Na^+] + [SO_4^{--}]$$

The square brackets used around the above chemical symbols are used in chemical technology as abbreviations for " the concentration of." Now there is a law of chemistry which states that the product of the concentrations of the ions will be constant under a given set of conditions. This product has been termed the solubility product, which may be represented graphically as $[A^+][B^-]$, or in the case of sodium sulphate as $2[Na^+][SO_4^{--}]$ or in the case of calcium carbonate as $[Ca^{++}][CO_3^{--}]$, and so on.

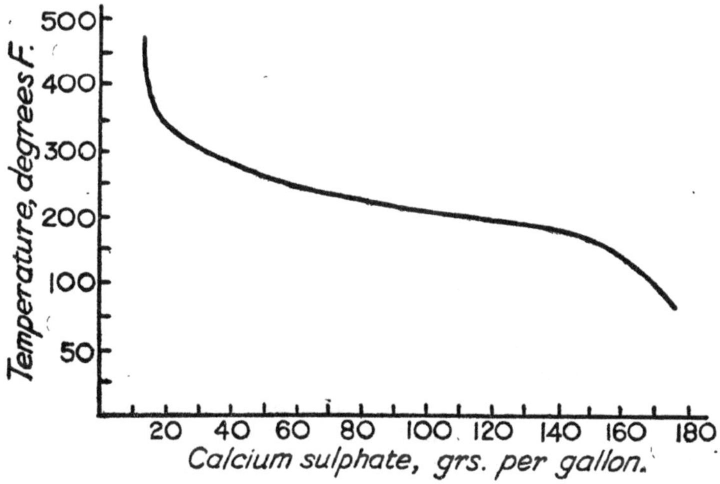

FIG. 11. Solubility of Calcium Sulphate.

Now if the concentration of one of the two ions *A* or *B* is increased by the addition of some other salt which on ionization produces either *A* or *B* ions, then some of the salt *AB* must go out of solution (i.e., some of the *A* ions and *B* ions must recombine to form non-ionized salt *AB*) so that the solubility product $[A^+][B^-]$ may remain constant. To take a definite case, consider a saturated solution of calcium carbonate to which sodium carbonate is added. The concentration of carbonate ions is now greater than with the calcium carbonate alone, and to maintain the solubility product of calcium carbonate constant some of the calcium ions must go out of solution. They can only do this by recombining with some carbonate ions. In other words some solid calcium carbonate is precipitated from the solution.

Briefly, this means that it is possible to adjust the solubility of a given salt in water by adjusting the concentration of one of its ions, and this is done by the addition of some other salt which also yields this ion on solution. This is very important from a scale prevention viewpoint, as will be appreciated if attention is given to the solubility relationships of two salts in the same saturated solution.

Suppose there are two salts *AB* and *AC* in solution, and that deposition or precipitation of the salt *AB* is preferred. Normally, on evaporation of such a solution, one of these salts will be deposited before the other; this will be the one whose solubility is first exceeded. As shown above, the solubility product of a salt can be increased by the addition of another salt and precipitation be so caused. Now in the solution of *AB* and *AC* it is the hard scale-former whose precipitation must be avoided, i.e., *AC*. Therefore a salt must be added which causes the solubility product of *AB* to be exceeded before that of *AC*. Such a salt will contain an ion which is found in *AB* but not in *AC*, e.g., a salt *BD*. Thus, in the case of a solution exactly saturated at one and the same time with calcium sulphate and calcium carbonate, the addition of sodium carbonate will cause precipitation of the calcium carbonate before that of the more troublesome calcium sulphate. The addition of sodium sulphate would cause the opposite effect by increasing the concentration of sulphate ions. In this case, therefore, the relative concentrations of carbonate (CO_3) ions and sulphate (SO_4) ions determine whether calcium is removed from solution as carbonate or sulphate.

These principles have been given practical value by the determination of the relative amounts of carbonate and sulphate, and the construction of CO_3/SO_4 ratio curves. These values are equilibrium values, and each point on the curve represents a solution saturated (at the temperature concerned) with both calcium carbonate and calcium sulphate. To ensure carbonate precipitation rather than sulphate deposition, sufficient carbonate must be maintained in the water to give a ratio slightly higher than the equilibrium value on the curve. The relative amounts of carbonate and sulphate are found by normal analysis. Originally, R. E. Hall[19, 24] calculated the necessary amount of carbonate from available solubility data, and plotted a curve of CO_3/SO_4 ratios. Naturally, the amount of carbonate required varies according to the boiler pressure (and temperature, therefore) since this affects the solubility, and also according to the amount of sulphate present. It has since been found that the presence of other salts in solution complicates affairs by modifying the solubilities of calcium carbonate and calcium sulphate. The work of Clark and Hunter[23] showed that sodium chloride, caustic soda and sodium sulphate all reduce the necessary ratio of CO_3/SO_4, the amount of reduction depending on the temperature and the amount of other salt added. Sodium sulphate was found to have the most pronounced effect in this way, an important fact when one considers that sodium sulphate is a constituent of most boiler waters and was originally thought to increase the necessary ratio value. It is neither feasible nor desirable in this book to reproduce details of the influences observed with these salts ; reference to the publication must be made for such data. Clark and Hunter's curve of CO_3/SO_4 ratios, obtained with a solution of calcium carbonate and calcium sulphate, is reproduced in Fig. 12 in combination with the original curve (shown as a broken line) published by Hall for comparative purposes.

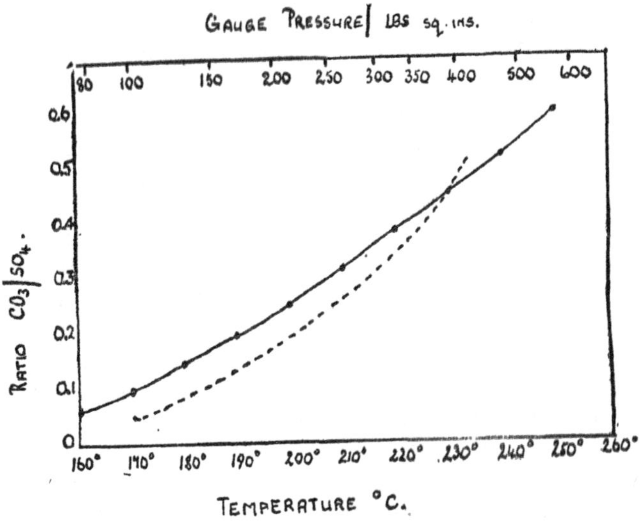

FIG. 12. Carbonate Conditioning Ratios.

In view of the discovery of the influence of other salts on the CO_3/SO_4 ratio, it seems desirable to combine control of carbonate conditioning, by means of carbonate and sulphate tests, with a determination of other salt ratios. In the blow-down tests in the Analysis section of this book, it is mentioned that saline concentration in the boiler can be watched conveniently by determining the ratio of chloride in the feed to chloride in the blow-down. Sulphate scale formation can be watched in the same way. The chloride ratio should be determined as described ; find also the ratio of sulphate in feed to sulphate in blow-down ; the ratio of sulphates will equal that of the chlorides if no sulphate scale is depositing, whereas if it is less then sulphate scale is depositing somewhere. A higher sulphate ratio will indicate removal of old sulphate scale. Such a periodic test, combined with a hardness test on the boiler blow-down, will indicate whether for any reason sulphate scale is forming (risk of scale formation is indicated by hardness in the blow-down water) due to ineffectiveness of the CO_3/SO_4 ratio being maintained. More carbonate can then be added in such a situation. Where necessary, the carbonate concentration can be increased by addition of soda to the feed tank or by using a slight excess of soda in the lime-soda softener.

At high temperatures and pressure a new factor enters which complicates the issue. This is the instability of sodium carbonate in aqueous solution. At temperatures above 185° C. sodium carbonate hydrolyses in water according to the equation

$$Na_2CO_3 + H_2O \rightleftharpoons 2\,NaOH + CO_2.$$

Removal of the carbon dioxide by the steam allows the reaction to continue in the right-hand direction, and considerable amounts of free caustic alkali may be produced.

Conditioning with sodium carbonate is therefore not feasible under just those conditions where conditioning is most important, viz., high pressures. The chemical sodium phosphate, Na_3PO_4, is therefore used instead, and this salt remains stable under the most drastic conditions. Instead of having calcium carbonate precipitated, there is tricalcium phosphate, which also does not form a hard scale.

In order that the calcium shall be thrown out of solution as the *tri*calcium phosphate, in which way the maximum amount of calcium is removed from solution by a given amount of phosphate, the water should possess a caustic alkalinity of about 100 p.p.m. expressed as hydroxyl (OH) ion. If this precaution is not adopted either or both the dicalcium phosphate, $Ca_2H_2(PO_4)_2$, and the monocalcium phosphate, $Ca(H_2PO_4)_2$, will be produced. These compounds, apart from the fact that they contain less calcium and therefore mean a smaller removal of calcium by a given amount of phosphate, are also much more soluble than the tricalcium phosphate. The requisite hydroxyl concentration is generally produced by the sodium phosphate, the reaction between this and the calcium and magnesium carbonates and hydroxides producing sodium carbonate, which decomposes as shown above, and caustic soda. If the hydroxyl concentration so produced is insufficient, the lime-soda plant can be so operated that sufficient caustic soda is present in the softened water.

A consequence of the above demand for alkalinity in the boiler water is that if the water be highly alkaline it may be unnecessary to use the trisodium phosphate—di- or mono-sodium phosphate may be used satisfactorily since they will react with the alkalinity of the water to give trisodium phosphate for reaction with the residual hardness of the water. This procedure will therefore be advantageous in cases where the boiler water alkalinity is already too high.

The quantity of phosphate present in the boiler water must be such that the solubility product of tricalcium phosphate, $3[Ca^{++}] \times 2[PO_4^{---}]$, is always reached before the solubility product of calcium sulphate, $[Ca^{++}]\,[SO_4^{--}]$. If the concentration of SO_4 ions is such that the solubility product of the phosphate is exceeded by $[Ca^{++}]\,[SO_4^{--}]$ then calcium sulphate will be deposited and scale formed in the boiler. If, on the other hand, sufficient sodium phosphate is added to the feed water such that the concentration of the PO_4 ions causes the solubility product $3[Ca^{++}] \times 2[PO_4^{---}]$ to be exceeded before $[Ca^{++}]\,[SO_4^{--}]$, then the calcium will be precipitated as calcium phosphate, which does not form a hard adherent scale. Again owing to the variation of solubilities with temperature, the value of the PO_4/SO_4 ratio varies with the operating pressure.

Original PO_4/SO_4 ratios published[19, 25] have been abandoned as a result of later research and plant experience. More recent research[26] indicates that solubilities (and therefore ratios) are so influenced by salts such as sodium sulphate and caustic soda, and the presence of solid phases, that recommendation of specific ratios of PO_4/SO_4 concentrations would be unwise. Instead, it is considered desirable that there should be sufficient concentration of soluble phosphate ion in the feed water as will maintain in the boiler a steady reserve, and so deal with any chance influx of calcium ions into the boiler system. An additional safeguard may be imposed by checking the efficiency of the phosphate treatment by means of the sulphate and chloride ratios as described in connection with carbonate conditioning.

In recent times a number of large high pressure boiler plants have been controlled in this way, with very good results. An indication of the effectiveness of the procedure will be obtained by considering the case of a 1,300 lb. boiler plant at Milwaukee. Data published by Anderson[27] have shown that phosphate conditioning was quite successful in this case. He found that by maintaining a PO_4/SO_4 ratio of not less than 0·14 in the presence of free OH no scale was formed even on the hottest sections of the boiler heating surface. Previously, although the make-up consisted of distillate, tube failures were caused by calcium sulphate scale formed owing to condenser leakage.

With regard to the method of supply of the sodium phosphate to the boilers, the existence of the two other calcium phosphates—monocalcium phosphate and di-calcium phosphates—may be recalled. The first of these forms a very porous scale, and the second a hard adherent scale. They are, therefore, dangerous both inside and outside the boiler. If trisodium phosphate is added to the feed from the lime-soda softener to remove the residual hardness, and the water does not possess caustic alkalinity, these salts may be formed. The mono-salt will deposit in the feed lines and economizer, with the result that the feed lines will rapidly choke up. The di-calcium salt may form scale even in the boilers themselves. It is therefore best under normal circumstances to introduce the reagent direct into the boilers by means of high pressure pumps.

The amount of phosphate used must be equivalent to the residual hardness of the feed together with sufficient excess to maintain the necessary excess of phosphate ion. Consequently, the lower the residual hardness or the more efficient the preliminary softening, the less the phosphate which will be required.

Use of Metaphosphate and Pyrophosphate

The introduction of sodium hexametaphosphate, better known as metaphosphate, and sodium pyrophosphate, has proved extremely useful in combating the above disadvantages, and at present metaphosphate particularly is in extensive use in this country and the U.S.A.[28]

The primary advantage of these compounds in boiler feed water treatment is their ability to react with the alkalinity to form the orthophosphate already mentioned in connection with scale prevention. Considering the metaphosphate, it will react with either water or alkali at temperatures over 212° F. to give the orthophosphate as follows:

$NaPO_3$	+	H_2O	=	NaH_2PO_4		
Sodium metaphosphate		Water		Acid orthophosphate		
NaH_2PO_4	+	$2\,NaOH$	=	Na_3PO_4	+	$2\,H_2O$
Acid orthophosphate		Caustic soda		Sodium orthophosphate		Water
$NaPO_3$	+	$2\,NaOH$	=	Na_3PO_4	+	H_2O
Sodium metaphosphate		Caustic soda		Sodium orthophosphate		Water

This means that since the water of a properly regulated boiler is always alkaline, sodium metaphosphate can be added to the feed line with the assurance that orthophosphate will be formed *inside the boiler,* just where it is needed to precipitate the calcium salts, and in addition excessive alkalinity will be prevented from building up there. Furthermore, the metaphosphate when dissolved in water does not react with the residual calcium in the feed, but forms a clear faintly alkaline solution. No deposits are thus formed in economizer or feed lines, and feed pipes are also kept free from corrosive attack by reason of the unimpaired alkalinity of the feed. With orthophosphate the reaction which often occurs in the feed lines reduces the alkalinity there and allows risk of corrosion. The fact is that sodium metaphosphate can form complex compounds with calcium salts, the calcium presumably entering as a complex salt, thus:

$2\,CaCl_2$	+	$Na_2\,(Na_4\,(P_6O_{18})$ (or $6\,NaPO_3$)	=	$Na_2\,(Ca_2P_6O_{18})$	+	$4\,NaCl$
Calcium chloride		Sodium metaphosphate		Complex phosphate		Sodium chloride

If excess of a solution of metaphosphate be added to a solution of calcium no precipitate is formed, and if sufficient metaphosphate is present the mixture will not answer the usual chemical tests for calcium ions. Even calcium oxalate (the most insoluble calcium compound known) can be dissolved in sodium metaphosphate, if the latter is used in excess. This complex formation explains why metaphosphate will dissolve calcium precipitates already formed in economizers, feed heaters, and feed lines generally.

In short, metaphosphate holds up residual calcium salts in the feed water and delays their precipitation until the boiler is reached, where the sludge formed can be more easily removed. The reduction in alkalinity reduces the amount of sodium sulphate required in the boiler water, since certain minimum sulphate/alkalinity ratios have to be maintained to prevent danger of embrittlement of boiler plates. This naturally means a smaller saline content in the water, and high salinity is one of the major

PLATE 6. PERMUTIT "PRECIPITATOR" PLANT

PLATE 7. PATERSON "ACCELATOR" PLANT
(By courtesy of Leicester and Warwick Electric Power Co.)

PLATE 8. DRY CHEMICAL FEEDER

PLATE 9. KENNICOTT BASE-EXCHANGE PLANT

causes of priming. The easier control of alkalinity, and consequent savings in blow-down, are particularly important for heat economy in plants which do not operate heat exchangers.

Similar advantages result from the use of pyrophosphate, since this also reacts with water or alkali as follows:

$Na_4P_2O_7$ Sodium pyrophosphate	+	H_2O Water	=	$2\,Na_2HPO_4$ Acid orthophosphate		
Na_2HPO_4 Acid orthophosphate	+	$NaOH$ Caustic soda	=	Na_3PO_4 Sodium orthophosphate	+	H_2O Water
$Na_4P_2O_7$ Sodium pyrophosphate	+	$2\,NaOH$ Caustic soda	=	$2\,Na_3PO_4$ Sodium orthophosphate	+	H_2O Water

This has less effect on alkalinity and is therefore useful for less severe operating conditions. This salt forms definitely alkaline solutions with water, and therefore gives adequate protection of the feed lines from

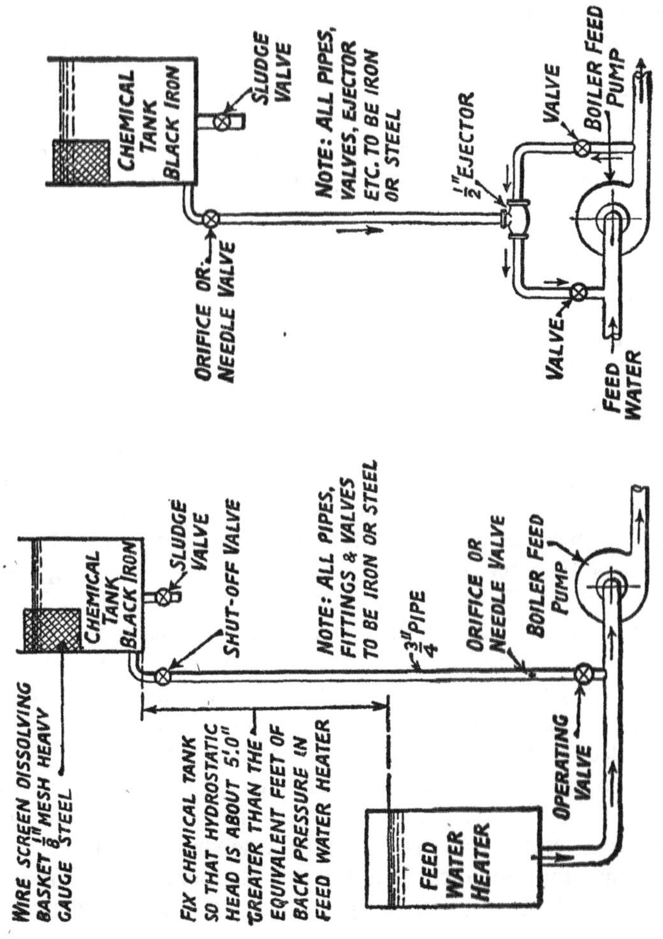

Figs. 13 and 14. Arrangements for feeding Sodium Metaphosphate.

corrosion. No deposits of calcium salts form in the economizers, feed lines, feed heaters, or anywhere before the boiler itself.

Meta- and pyro-phosphates, since they can be added quite safely to the feed lines, can be quite simply fed under gravity head to the suction line of the feed pumps. Even if slight deposition occurs in spots where the solution is subject to prolonged heating, as in economizers, the addition of a little cutch, quebracho, or similar substance to the feed (say 1 or 2 p.p.m.) will retard the conversion to orthophosphate (which is the cause of the deposit) until the boilers are reached. Methods of feeding metaphosphate are sketched in Figs. 13 and 14.

An investigation of the comparative values of metaphosphate and pyrophosphate, with the conclusion in favour of metaphosphate, has been published recently[29] which also contains data on the ability of metaphosphate to form complexes with calcium.

In recent years, phosphoric acid has been increasingly used in treatment plants. It will reduce the alkalinity even more than the above sodium salts, as shown by the reaction:

H_3PO_4	+	3 NaOH	=	Na_3PO_4	+	3 H_2O
Phosphoric acid		Caustic soda		Trisodium phosphate (orthophosphate)		Water

This method has received wide application, particularly in the U.S.A., either in a single stage or following lime-soda treatment as the second part of a two-stage hot-process softening under pressure.[3, 4] Effluent approaching zero hardness can be obtained by this procedure, and sludge accumulation in the boilers is therefore reduced to very small amounts. This is particularly important for plant operating above 1,000 lb. and it is interesting to note that successful use of this reagent has been reported for boilers operating at 1,400 lb. pressure.[30]

Budenheim Phosphate Treatment

A method of treatment with trisodium phosphate has been developed on the Continent, and is generally known as the "Budenheim Process."[31] This process, which has been widely adopted in Germany in particular, is designed to produce a completely softened and de-aerated water, and furthermore neither feed-line, etc., deposits nor boiler sludge are produced. Oil removal from steam may also be accomplished, the only proviso being that in such cases a larger filter capacity is required. The process consists essentially of three stages. A cascade preheater drives off dissolved oxygen and carbon dioxide from the raw water, and at the same time some part of the temporary hardness in the water is removed by thermal decomposition of bicarbonates. Incidentally, oily steam may be used for preheating, since the phosphate sludges finally removed by filter retain the oily particles and globules. If the curve for the solubility of oxygen with temperature is consulted elsewhere in this book, it will be realized that maintenance of a suitable temperature is important for complete removal of gases in the preheater. In the second stage of the process the hot water from the preheater is mixed with alkaline water returned from the boiler. Partial softening of the water is effected by the alkalis in the boiler water, and this mixed and partially softened water then passes to the third stage, where it receives treatment with trisodium phosphate. In this stage, speed of softening is assisted by the raised temperature resulting from the cascade preheating and mixture with hot

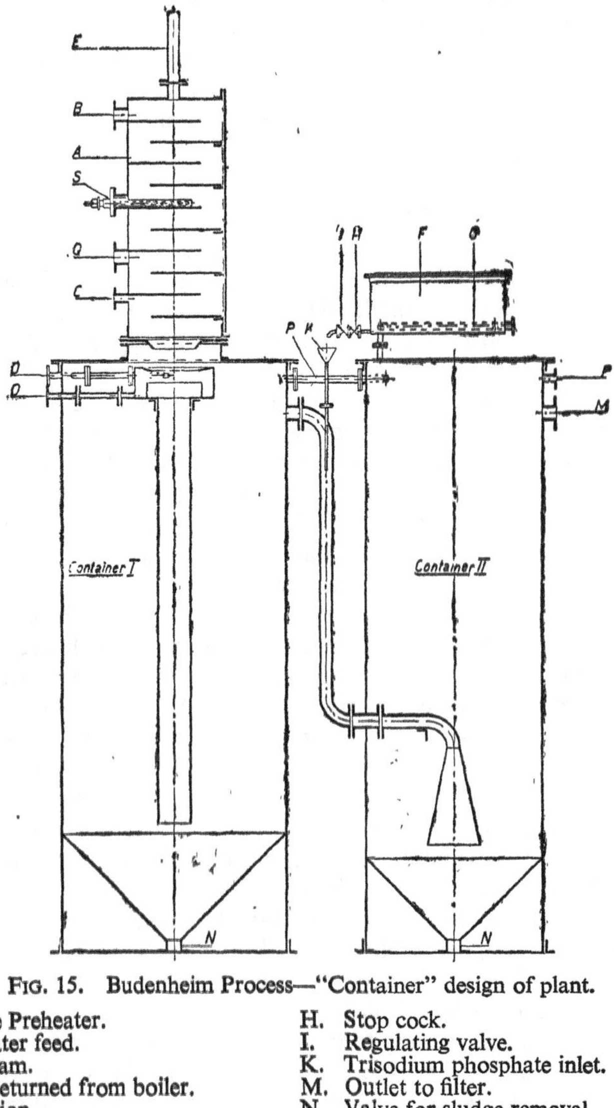

FIG. 15. Budenheim Process—"Container" design of plant.

A. Cascade Preheater.
B. Raw water feed.
C. Live steam.
D. Water returned from boiler.
E. Ventilation.
F. Trisodium phosphate container.
G. Heating coil.
H. Stop cock.
I. Regulating valve.
K. Trisodium phosphate inlet.
M. Outlet to filter.
N. Valve for sludge removal.
O. Outlet for boiler water.
P. Steam channel.
Q. Waste steam inlet.
S. Thermostat.

boiler water. Softening is claimed to be completed practically to zero hardness in this stage, the precipitated sludges being removed in a suitable filter. Under such conditions, of course, no sludge can enter the boiler and assist development of priming troubles. For the same reason, deposits in feed-lines, economizers, etc., are eliminated. The flocculent phosphate precipitate produced in the final stage is said to deal satisfactorily with oily matter, finely divided suspended matter, etc.

Two main types of plant have been developed for applying the Budenheim Process. The so-called " container " plant is illustrated in Fig. 15. A tubular design is a recent development for application to conditions, such as on board ship, where confined space calls for a compact form of plant. Both types of plant produce the same result.

Silica Scales and Deposits

Silica is objectionable in boiler water because it may lead to the formation of hard, dense silicate scales having high resistance to heat transfer, because turbine blade deposits containing silica are difficult to remove, and because it may assist in embrittling the boiler steel. The analcite type of complex silicate scale was first reported by Powell in 1933, but its deposition in boilers has increased in recent years owing to the trend towards higher boiler pressures. This agrees with research by Partridge and Purdy[32] on the mechanism of formation of this type of scale, as a result of which they concluded that only above 400 lb. pressure (more usually above 600 lb.) were operating conditions such as heat transfer rates usually suitable for complex silicate formation. Simpler silicates may deposit at lower pressures.

All natural waters contain silica in some form, largely as a result of the weathering of rocks, the silica content varying in different localities but in general tending to be higher in soft waters. It has been found[33] that the nature of the silica in water depends to a great extent upon the other minerals present, and to a great extent upon the hydrogen ion concentration. Both colloidal and crystalloidal (soluble) forms exist in water, natural supplies usually containing ¼ to 1 grain per gal. A maximum silica concentration of 12 p.p.m. in the boiler water is commonly adopted as the operating limit, but even lower limits are required at very high pressures. These different forms in which the silica exists, and the very small amounts involved, have led to many different removal treatments being proposed. However, although considerable research on the problem has been undertaken in recent years, plant experience with the various treatments is still fairly limited.

A number of investigations of treatment by chemical coagulation have emphasized the important rôle of adsorption in the removal of silica. German work[34] with sodium, barium, and calcium aluminates as reagents led to the conclusion that the mechanism of silica removal was largely one of adsorption. The efficacy of sodium aluminate was confirmed by Lindsay and Ryznar,[35] who found that an optimum pH of 8·5 and recirculation of precipitated sludge increases the silica removal; these workers, too, concluded that silica was removed mainly by adsorption. Further evidence of the major part played by adsorption was obtained by Schwartz[36] when studying the use of ferric sulphate and hydrated ferrous oxide. Again the importance of an alkaline pH was revealed, since excellent silica removal could be obtained when the water was treated with ferric sulphate and lime to a pH of about 9·0, whereas little or no silica removal resulted with coagulation at an acid pH. Christman *et al.*[37] removed silica by using magnesium sulphate and sodium aluminate, which they found gave a flocculent adsorbent precipitate of magnesium aluminate, with some formation of aluminosilicate floc also under the correct molecular proportions. Residual soluble alumina must be present. Straub[38] obtained good silica removal by using excess magnesium oxide at temperatures from 182° C. to 282° C. and subsequently filtering the treated water. Geisler[39] reported that at a high-pressure plant in Germany silica was removed with magnesium oxide and caustic soda. Normally the feed contained 4–5 mg. silica per litre, which concentrated in the boiler water to about 70 mg. per litre. Filtration of the feed after softening through a filter of granular magnesium oxide gave good results, but the filter was soon exhausted. Treatment with magnesium oxide and caustic soda was tried over a period of a year, the process reducing the boiler water silica content to 18 mg. per litre (about 1¼ grains per gal.). These early experiences have led to the development of several methods of treatment.

In the Betz adsorption process,[40] removal of silica is effected by a specially prepared form of magnesium oxide, and recirculation of partially spent magnesia sludge is also utilized to reduce the quantity of fresh magnesia and reduce the final silica content of the water. It is claimed that the process is well adapted for use in conjunction with hot-process lime-soda softening, and requires no increase of lime and soda dosage; best results are obtained at a pH of about 10. The efficiency of removal is stated to increase with temperature, the major increase in the percentage of silica removal occurring with an increase in temperature up to 95° C. (203° F.), the approximate temperature of operation of conventional hot lime-soda softeners. Furthermore, where the silica content of the water is to be reduced to 2–5 p.p.m., little is gained by increasing the time of retention beyond 1 hour at this temperature. Equilibrium is said to be almost complete in about 15 min., in which time about 85 per cent of the silica is removed, but in the cold a much longer time is required. An important advantage of the adsorption technique is that reduction of soluble silica (usually the major part of the silica present in waters is soluble, or crystalloidal) is achieved without increase in the solids content of the treated water, and in fact some decrease.

Since the removal of silica by chemical coagulation is largely a matter of adsorption, the efficiency with which the silica is removed will be greatly affected by the intimacy of contact secured when the coagulant and silica-bearing water are mixed. H. L. Tiger[40] has described the application of a modified Spaulding Precipitator plant to provide this intimate contact in an improved magnesia process which combines the use of ionic magnesium, precipitated as $Mg(OH)_2$, with magnesia sludge. The equipment permits the usual lime-soda reactions to take place at the same time as the silica removal and it is stated that superior silica removal is obtained by the positive agitation and intimate sludge contact in this type of plant. Ionic magnesium can be added to the raw water economically without increasing the lime requirement or the total solids content of the effluent water by recirculating sludge into a magnesium dissolver compartment, or low pH zone, where bicarbonate ions and dissolved CO_2 in the raw water dissolve magnesium ions (Mg^{++}) from this sludge. If necessary, additional carbon dioxide can be introduced to the raw water to increase the amount of ionic magnesium so introduced. Tiger confirms that the efficiency of silica removal increases with temperature, but claims satisfactory results from this process at low temperatures (85°–100° F.) owing to the greater efficiency of ionic magnesium, improved contact from sludge agitation, etc. It is found that the presence of phosphate ion interferes with silica adsorption by magnesia, and consequently it is advised that any phosphate treatment should follow the silica treatment.

Another magnesia process is the British "M.W." treatment,[41] which makes use of the boiler itself as a reaction vessel and thus secures prolonged and intimate contact at the high temperatures favourable for best removal of silica. A milk or cream of specially prepared magnesium oxide is introduced direct to the boiler water either while the boiler is off range or by means of an injector whilst steaming.

Various other treatments have also been used with some success. Iron compounds, particularly ferric and ferrous sulphate, as mentioned above, have been found quite effective. For instance, Powell claimed[42] that the silica content could be reduced to 3 p.p.m. or less by the addition of ferric sulphate and caustic soda (which, of course, would react to produce the coagulant ferric hydroxide). Ferric sulphate is quite commonly used in cold process softening, but has the disadvantage that the sodium sulphate produced increases the total solids content. Phosphate conditioning may also be helpful against silicate scaling, since the phosphate combines with the residual calcium in the softened feed to form calcium phosphate instead of silicate. Some formation of calcium silicate must occur, however, owing to the sharing of calcium in the proportions of the respective solubilities of the two compounds. At high temperatures results will not be satisfactory since such temperatures will favour back reaction between calcium phosphate and the silica. The addition of tannins improves results, but it is inadvisable to rely upon this method above 200 lb. pressure. Other materials such as organic glucosides[43] have also been recommended but not widely used. The action of all these compounds has been either to maintain the silica in solution or to precipitate it as a non-adherent scale. The effect depends upon the operating conditions, especially with respect to boiler pressure and heat transfer rates, and they are probably more suitable for relatively low operating pressures.

Turbine Blade Deposits

Scaling of turbine blades has caused increasing trouble in recent years, particularly in the larger electrical generating stations. The serious effect of such deposition is evident from cases which have been quoted where the efficiency of machines dropped something of the order of 20 per cent after only a few weeks' service. The turbines then had to be allowed to cool down for removal of the deposition, and when such idle periods are added to the gradual loss of efficiency and output the matter becomes very serious on a large plant.

The deposits are produced, of course, from mineral matter carried in the steam, and may be either soluble or insoluble, the latter type usually containing a high proportion of silica. The soluble deposits can be removed by washing with water, but caustic soda solution is required to remove the insoluble silica type. Sometimes the soluble deposit contains an insoluble component, in which case after frequent washings the accumulation of insoluble material makes water washing ineffective. A number of analyses of typical deposits from turbines operating under various conditions is given below[44]:

TABLE 5

Chemical Analyses of Turbine Blade Deposits.
(Composition in per cent)

Steam pressure lb./sq. in.	400	600	600	600	800	1200	1200
Silica SiO_2	3·2	80·1	81·3	1·4	9·3	85·2	53·9
Alumina Al_2O_3 ..	5·8	2·1	1·9	6·3	2·0	—	—
Iron oxide Fe_2O_3 ..	1·8	1·0	2·0	4·0	1·2	3·6	24·6
Sodium oxide Na_2O	27·3	3·6	10·2	33·4	56·3	2·5	5·6
Calcium oxide CaO ..	16·8	0·2	trace	trace	1·9	trace	3·9
Magnesium oxide MgO	8·2	0·2	—	—	0·5	—	—
Chloride Cl ..	—	6·5	trace	—	10·1	—	—
Phosphate P_2O_5 ..	—	trace	—	17·3	9·6	1·7	7·3
Carbonate CO_2 ..	36·3	0·5	0·5	1·2	9·2	trace	—
Sulphate SO_4 ..	—	trace	trace	1·7	—	—	—
Type of deposit ..	Soluble	Insoluble	Insol.	Soluble	Soluble	Insoluble	Insol.

Probably the first systematic examination of the problem was an investigation at the University of Illinois Experimental Engineering Station,[45] in which data were collected from twenty-six generating stations as a basis for the research work. In general it was found from such records that: (*a*) The deposits were not limited to any one temperature or pressure; (*b*) the deposits were not proportional to the total concentration of solids in the boiler water, and in fact deposits were obtained in which the relative proportions of the constituents bore no relationship to the ratios in which they were present in the boiler water; (c) the total solids in the steam were very low in many plants where the deposits occurred, and a steam which would normally be referred to as one of excellent quality might cause a great deal of deposit; (d) the deposits all contained sodium present as hydroxide, carbonate, or silicate: the chemical causing the major part of the difficulty appeared to be the caustic soda, which, of course, is present in practically all boiler waters; (e) the deposits were largely soluble in water. Straub concluded from these experiences that the fouling was due to the chemicals carried in the steam, and that the adhesion of the chemicals was due to the action of caustic soda as a sticky binder. When a solution of sodium hydroxide or caustic soda is subjected to heat it merely loses water to become more concentrated, and after reaching a certain con-

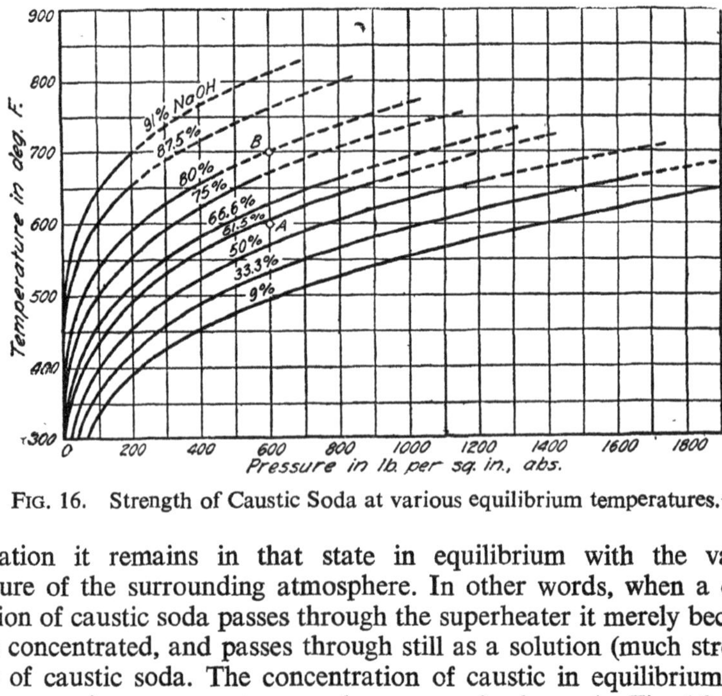

FIG. 16. Strength of Caustic Soda at various equilibrium temperatures.

centration it remains in that state in equilibrium with the vapour pressure of the surrounding atmosphere. In other words, when a dilute solution of caustic soda passes through the superheater it merely becomes more concentrated, and passes through still as a solution (much stronger now) of caustic soda. The concentration of caustic in equilibrium with steam at various temperatures and pressures is shown in Fig. 16, from which it will be seen that at the higher pressures prevailing to-day caustic solutions of 80–90 per cent strength may pass to the turbines. Such a solution will be present in steam as a thick paste or semi-fluid, which will readily adhere to the turbine blades. It will be appreciated that other salts present as a fine dust can readily adhere to the initial sticky deposit, and in this way a troublesome deposit may build up on the blading.

This was tested in a small experimental boiler with individual salts and mixtures such as are encountered in boiler practice. It was thoroughly confirmed, and, in fact, actual visual observation of the stationary blade employed showed that the caustic soda had caused the deposit by its being in a fluid, sticky state.

Preventive measures were proposed by Straub as a result of further laboratory experiments and trials on the turbines of a central station operating at 600 lb. pressure. The control proposed was the provision of other chemicals to dry up or reduce the stickiness of the caustic soda, and it was found that a ratio of sodium sulphate/caustic soda not less than 4·4 prevented deposition. To overcome the disadvantage of increase in salinity involved, the effect of other compounds normally present in boiler water was examined. Reduced deposition, but not complete prevention, resulted with a ratio of sodium sulphate/caustic soda at 3·3 combined with a ratio of sodium chloride/caustic soda at 2·5. Various organic materials were also tested, and limited laboratory data gave evidence of superior performance. A ratio of pyrogallol/caustic soda at only 2·2, or gallic acid/caustic soda at 1·86, proved effective, but detailed plant trials were not done. It should be noted that these materials were probably effective owing to their tendency to decompose to carbon dioxide, which would form sodium carbonate with the caustic soda and thus " dry " it chemically rather than mechanically.

Successful inhibition of deposits by these methods has been reported by Fellows[46] and Tray,[44] the latter stating that examination of a large number of blade deposits confirms the sodium hydroxide theory. German investigations[47] also led to the conclusion that the main agent responsible for the occurrence of soluble deposits is caustic soda, and that to control the insoluble silica deposits the amount of silica in the boiler water must be limited to 5–10 p.p.m. Operating details of a few German plants[47, 48] are reproduced below:

TABLE 6

Turbine Blade Fouling in German Plants.

Boiler pressure lb./sq. in.	966	1,704	1,704	1,896
Salt content				
in steam, p.p.m.	2–3	2–3	—	3
in boiler water				
total p.p.m.	2,000–3,000	3,000	1,500	2,500
NaOH p.p.m.	200–250	30	20–25	60
SiO_2 p.p.m.	30–50	15	25–30	45
Type of deposit	Soluble	Insoluble	Insoluble	Insoluble
Decrease in turbine capacity, per cent	10	10	10	10
Hours operation	1,200	4,000	260	1,500

Silica in the form of silicic acid in the steam has been believed to cause scaling of turbine blades, particularly the high silica deposits which have been experienced on an increasing scale in recent years as operating pressures have been taken higher. Past attempts to determine the volatility of silicic acid have been unsatisfactory, but Splittgerber recently reported experimental data[49] which showed that silica could be vapourized from silicic acid in appreciable amounts at pressures in the region of 1,400 lb. This work was taken further by Straub and

Grabowski[50] in laboratory research followed by power plant tests at around 1,250 lb. pressure. Those tests indicated that silica can leave the boiler as vapourized silicic acid, which later crystallizes on the blades in the lower pressure stages of the turbine. The plant tests indicated that when the silica in the boiler water is kept below 5 p.p.m. and the pH of the water around 11, no appreciable silica deposits form in the turbines. It was found that when the silica in the steam became greater than 0·1 p.p.m. deposition occurred, so that the silica content of the steam must be kept low either by scrubbing from the steam before superheating or by keeping the silica content of the boiler water very low. It will be recalled that a comparable boiler water figure was proposed by the German workers,[47] who also proposed a limiting caustic soda content of 10–25 p.p.m. to minimize fouling by soluble deposits. Recent German practice has been to limit residual silica in the feed to 1 p.p.m., and boiler water silica to less than 7 p.p.m. for high-pressure plant. It is of interest to note that at these high pressures (1,000–3,000 lb./sq. in.) an important and intricate investigation by Spillner[51] has shown that steam at these pressures can dissolve measurable amounts of compounds such as NaCl, NaOH, and KCl. His work so far has been limited to single salts in steam rather than mixtures such as occur in boiler water, but the results offer an insight into the mechanism of possible future steam contamination problems at high pressures.

Removal of water-soluble deposition from turbines is commonly effected by reducing the inlet steam temperature sufficiently to produce wet steam in the stages where the washing action is required. Periodical analyses of the hotwell condensate while washing determine the character of the deposit being removed, and also show when no more is being removed by the washing. Until recently, the usual practice for removal of insoluble deposits necessitated disassembly of the unit and mechanical removal from the affected surfaces with fly ash, etc. More rapid cleaning and reduced labour charges result from the use of caustic soda solutions as first tried in Germany by Geisler[39] and Goerke.[47] They sprayed a 10–20 per cent solution into the steam fed to the turbine and allowed the solution to remain in the machine for 3–4 hours. Subsequent water-washing was claimed to have removed all deposition. This method has been successfully used in the U.S.A.[52] to remove deposits containing 80–89 per cent silica from low-pressure turbines on a 1,350 lb./sq. in. plant, 10 per cent caustic being used for a 3-hour treatment period, followed by a 3-hour washing period using condensate to clear the system of caustic alkali.

Potassium Treatment

A new line of attack on boiler water problems was developed by R. E. Hall in 1942. Briefly, this consists of the substitution of sodium by potassium in the chemical equilibrium by using potassium salts in treating the feed water instead of the more common sodium compounds. Cost is the chief objection, since sodium compounds are easier and cheaper to obtain, but with expensive very high-pressure plant this factor becomes less important if superior results are achieved.

Although chemically sodium and potassium are very similar elements, Hall[53, 54] has emphasized certain notable differences in certain of their compounds and proposes that advantage be taken of these differences to combat various water difficulties which have become increasingly acute as modern power plant pressures advance. Many potassium salts are more soluble in water than the corresponding sodium salts; for example, in the boiler water temperature range potassium silicates are increasingly soluble as the temperature increases, whereas sodium metasilicate and disilicate decrease in solubility with rising temperature. There is therefore less silicate deposition from the concentrating film of boiler water at steam bubbles (see page 57) with the potassium equilibrium. It is suggested that this may lead to reduction in siliceous scaling generally and in addition assist boiler control of magnesium by allowing the presence of enough silica in the boiler water to result in removal of non-adherent magnesium silicate. Improved turbine conditions are also claimed since potassium silicates are more stable and it is suggested that at the alkalinities generally carried in boilers the sodium silicate in the carry-over is in a form that will precipitate silica more readily in certain stages of the turbine than will the corresponding potassium silicates. The latter are also said to be removed easily by washing with wet steam whereas the former will require alkali washing. A further advantage of potassium treatment is claimed; owing to the greater solubility of potassium salts, "hide-out" of treatment chemicals is greatly reduced.

Although the potassium treatment has been tried on a number of operating boilers,[55] at pressures ranging from 1,350 lb. to 1,950 lb., more laboratory and field data are required to reveal the full technical characteristics of the treatment and enable its value to be accurately assessed. The experience to date has been good, but readers who are interested are advised to consult the original papers quoted for details.

Chlorination of Condenser Cooling Water

The slimy gelatinous lining which generally forms on condenser tube surfaces and reduces cooling efficiency is due to growths of micro-organisms known generally as algae, which are almost invariably present in cooling water supplies. Condenser conditions are particularly favourable for the development of these organisms, since their growth increases particularly as the temperature rises to 60°–100° F. When chlorine is admitted to the cooling water, their growth and development is stopped and no slime is produced to enable mud and suspended matter to bind together and adhere to the tube surfaces.

The chlorination of cooling waters has been applied both as a continuous and as an intermittent process. There are also two methods of introducing the chlorine. These are as a gas and as a solid—calcium hypochlorite. Liquid chlorine is generally supplied by chemical firms in steel cylinders at a pressure which causes the gas to liquefy, but upon discharge from the cylinder it immediately gasifies. Calcium hypochlorite is the older of the two forms and is still preferred by some users, more particularly where the quantity of water to be treated is comparatively small. Chlorine gas may be introduced to the water by means of a vacuum chlorinator. In one outfit, a vacuum is produced by an auxiliary water supply flowing through an injector and the chlorine is thus drawn into the injector and so becomes mixed with the auxiliary water supply. This forms a chlorine solution, which is mixed with the main cooling water supply. Arrangements are made so that if the water supply to the apparatus fails for any reason, then the feed of chlorine automatically stops. Other safety measures are also introduced, such as keeping the chlorine under vacuum as long as it is in the apparatus ; this reduces to a minimum the possibility of leakage of chlorine and possible danger to men working in the neighbourhood. Easy inspection by visual examination of the working parts is also generally arranged so that irregularities in operation can soon be detected. Apparatus for feeding calcium hypochlorite is similar in some respects to a lime-soda softening plant. The dry chemical is placed in a revolving drum which also contains the requisite amount of water. The creamy solution so formed is discharged to a main dilution tank to allow settlement of insoluble lime, which is ineffective in this process, and a clear chlorine solution is obtained for use as required. It should be mentioned that calcium hypochlorite contains only about one-third of its weight of available chlorine, so that large amounts are required for a given plant. Good stability of the hypochlorite is also a feature to be desired when purchasing.

The intermittent process of chlorination is, of course, a modification of the continuous process, the chief advantage for cooling water supplies being a reduction in the cost of chlorine while highly efficient results are still obtained. This reduction is particularly important with cooling water supplies, which are often very large in quantity. The time interval for introducing the chlorine is regulated to suit the water and the operating period in some cases has been as low as three minutes during each hour of plant operation. This intermittent feeding is quite mechanical, a clock making contact for a relay which operates a solenoid valve in the water line to the injector, when the vacuum caused by the flow of water draws the chlorine in through the reducing valve, etc.

An important requirement of any chlorination outfit is that the apparatus be constructed of materials which will withstand the corrosive powers of the gas. An essential feature of such protection will be means for preventing access of moisture, particularly to delicate parts. Dry chlorine gas does not corrode at normal temperatures, but in the presence of moisture a particularly virulent attack is set up. With regard to the point of application of the chlorine, though much depends upon the lay-out of piping, it is advisable to introduce the chlorine to the water as near the intake as possible. This will give full dilution before the water enters the condenser.

The frequency with which it is necessary to introduce chlorine will depend entirely upon the quality of the water to be treated, and also its temperature. Thus, at the Kearny station in America some years ago,[56] it was found that the amount of chlorine required to treat a condenser was as follows:

TABLE 7

Chlorine Consumption Variation with Water Temperatures

River water temperature, °F.	Pounds per day	Number of days	Pounds of chlorine for period
32–50	40	121	4,840
50–70	60	152	9,120
70–80	75	92	6,900
		365	20,860

Similar variations with seasonal differences in water temperature are experienced by modern power stations in this country.[57]

When a condenser is operating normally with clean tubes a certain known temperature rise may be anticipated with a given installation. The rate of chlorine feed should be such that this temperature difference is maintained and not allowed to rise owing to fouling of the tubes. During the period of operation of the chlorinator samples of the cooling water should be taken every two minutes and tested for residual chlorine content, as described in the analytical section of this book. This residual chlorine should be kept as low as is consistent with the results to be achieved; generally a residual of 0·2 p.p.m. to 0·05 p.p.m. is sufficient. It should not be overlooked that the residual chlorine will vary with the time from which the chlorinator has cut in. Therefore, for the purpose of fixing the correct treatment, tests should be taken every two minutes during the complete period of the chlorinator's operation, and continued as the time-lag indicates.

The outstanding advantage of chlorination is the gain in vacuum which results (e.g., Kearny ¼ in., a large power station in this country treating 72,000,000 gal. of water daily—1½ in.),[58] since a variation of 1 in. affects the steam consumption and the fuel bill by something of the order of 6 per cent. In addition there is a considerable reduction in cleaning charges, with accompanying reduced outage periods for the plant.

One of the main disadvantages cited against chlorination is the fact that chlorine is poisonous and there is therefore thought to be some risk to workmen, etc. Actually, this is very slight, since all the well-known makes of chlorination apparatus pay special regard to this aspect and provide adequate safeguards, which are further implemented by the chlorinator house being placed at some distance from other buildings.

In any case, the choking effects of small amounts of chlorine gas will soon give warning of any trouble, when suitable measures can be taken. As regards corrosion, outfits are made to withstand this, and when chlorine is in weak solution in the circulating water, little attack on the condenser material need be feared. In conclusion, therefore, such treatment is well worth adoption, whatever the source of the water, both for condensers and cooling towers.

Another method of chlorination which has been successful is to admit chlorine gas into the condensers after shutting down and draining. After a period of contact with the wet surfaces, the gas is cleared by the circulating pump. In one case,[59] adopting a contact time of 15–30 min., growth of slime was prevented with 2 lb. of chlorine per 100 cu. ft. of space.

A comprehensive paper on the chlorination of cooling water has recently been published, and may well be consulted[60] for a full history of the subject as well as a description of present-day methods from engineering, chemical, biological, and economic aspects.

Ammonia is sometimes used in conjunction with chlorine, since it is claimed that smaller amounts of chlorine need to be used owing to the greater killing effects of the chloramine compounds formed by union of the ammonia and chlorine in dilute solution.

The toxic properties of chlorine are also used in the application of a new reagent[61]—sodium pentachlorphenate—originally developed in the U.S.A. This new substance can be added to the cooling water very simply in the form of briquettes, which slowly dissolve, and its application is available for any recirculating system or under conditions where the overflow or purge loss of water is not too great.

This new industrial algaecide has the following properties. Without being dangerous to man, it is highly toxic to organic growths and micro-organisms such as algae, protozoa, capsulated bacteria, etc. Usually a dosage of 15–20 p.p.m. is sufficient in cooling-water treatment, a somewhat higher dosage (30 p.p.m.) serving to kill capsulated bacteria where this is necessary. This algaecidal power is maintained uniformly throughout the usual hydrogen-ion concentration range experienced in cooling-water practice. The reagent is readily soluble in water and is stable, and the chlorine remains unaffected by oxidation, aeration, or other minerals or conditions to which it is likely to be exposed in cooling systems. It does not decompose to develop acidic solutions, and in fact is non-corrosive to the metals used in engineering construction work, particularly under the concentrations normally used for algae control. The concentration in water does not decrease by continued application of heat. It follows that little of the reagent is lost from the cooling system other than by purge, etc. A final important point is that this treatment does not interfere with other treatments which may be given to the cooling water, such as for the prevention of scale or corrosion. It has been found advisable, however, if this chemical is used in conjunction with the well-known scale-preventive sodium metaphosphate, to add the pentachlorphenate first since a mixture of the two chemicals in high concentrations will tend to precipitate pentachlorphenol and convert the metaphosphate to orthophosphate as follows:

CCl_5ONa	+	$Na PO_3$	+	H_2O	→	CCl_5OH	+	Na_2HPO_4
Sodium pentachlor-phenate		Sodium metaphosphate		Water		Pentachlor-phenol		Disodium phosphate

It is not difficult to develop the most suitable operating procedure for individual plant conditions. For clearing water systems already fouled with slime and algae deposition, the following general procedure can be adopted. First, a concentration of about 30 p.p.m. of the chemical should be established in the water system. This concentration should be maintained for about ten to fifteen days, this period usually being long enough to kill and remove the existing deposits. The algaecidal action can be watched by noting the change in colour of the growths from greenish to brown or grey. The chemical concentration is then gradually reduced over a period of two weeks or so by decrements of 5 p.p.m. until the algae growth just begins to reappear. The concentration is then increased by about 5 p.p.m. and maintained at that level by the regular addition of sodium pentachlorphenate. This concentration will then usually have sufficient toxic value continuously to inhibit any growth. Analytical methods for the determination of the sodium pentachlorphenate concentration are available to assist the user to keep the concentration in use at a stable level.

The application of sodium pentachlorphenate to cooling-water systems requires no complicated equipment, and no weighing, measuring, or proportioning apparatus is required for its application, although standard types of wet or dry feed mechanisms may be used if desired. The material can be handled safely, and requires no special loading or storage facilities. It is supplied in 1 oz. tablets or briquettes which may be allowed to dissolve slowly in a wire basket hung in a circulating-water passage. Alternatively, wire or cords may be tied to the bags of briquettes and the whole lot (bags and all) then immersed in the water. The briquettes will usually dissolve in two to five hours, according to the water flow. The exact dosage of sodium pentachlorphenate varies, depending upon the organisms, the extent of contamination, and environmental factors. Environmental factors are important. For example, during summer when water temperatures are higher, growth will be correspondingly increased, and larger dosage will then be necessary to check it. It is a simple matter to determine the correct dosage by experiment, however.

Since sodium pentachlorphenate is not volatile nor highly reactive chemically, it is not lost by evaporation or precipitation. Practically the only loss which must be replaced is that resulting from windage, leakage, and purge from the system. These daily losses are compensated by a single make-up charge added each day, and is based on the daily addition of make-up water. The initial charge is calculated from the static capacity of the system. Thus, 100,000 gal. of water will require an initial dose of 30 lb. of pentachlorphenate to allow a dosage of 30 p.p.m. in the water. The principal cost of the treatment is the initial cost of the chemical, which varies with the size of the system and the extent of the loss of water. There is no cost for equipment, and labour charges are very low, so that the operating cost depends largely on the daily make-up water quantity. The cost for local conditions can therefore be calculated from the cost of treating the daily overflow loss ; in addition, of course, there is the initial cost of treating the full water capacity of the cooling system. On one large system, of 10,000,000 gal. capacity, the cost of treating the make-up water amounted to less than £100 for the first year, applying a treatment of 15 p.p.m. In general, operating experience has shown that in most cases 10–20 p.p.m. of sodium pentachlorphenate will give good control of slime and algae formations, while in addition growth of crenothrix and other so-called " iron bacteria " can also be controlled by its use.

A final note on the human aspect will be appropriate. Chemicals which are sufficiently toxic to micro-organisms to suppress their growth effectively might be expected to be harmful to the human system. Careful toxicological studies have shown that the material is not highly toxic to man, and the fact that only low concentrations are used in algae control is a further safety factor. It is recommended, however, that cooling water containing sodium pentachlorphenate should not be used for drinking or bathing purposes. The explanation of the latter caution is that although, normally, solutions containing not more than 100 p.p.m. of the substance are non-irritant, even on long exposure, the skins of individuals vary somewhat. For this reason, it is worth while washing the hands with soap and water after handling in order to avoid any possibility of mild skin irritation. In many instances, it will be possible to immerse the bag of reagent (bag and all) in the water, thus minimizing handling. Should it be necessary to use less material, however, a number of the non-dusting briquettes can be removed and placed in another bag or a wire cage for solution.

Scale Formation in Cooling Water Systems

In addition to formation of organic growths and slimes on cooling surfaces, scaling and corrosion troubles may be experienced due to the mineral content of the cooling waters. Cooling water supplies often contain substantial amounts of such impurities, since they have to be chosen for cheapness rather than purity.

The scaling of condensers is a different problem from that of steam boilers. Deposits can form on boiler heating surfaces both from the carbonate and non-carbonate hardness, but in the case of a cooling water system any scale deposits are usually formed from the carbonate

PLATE 10. ELECTRICALLY OPERATED AUTOMATIC BASE EXCHANGE WATER SOFTENER
(*Boiler House Review*)

PLATE 11. PERMUTIT AUTOMATIC CONTROL VALVE

hardness; calcium carbonate particularly forms by far the major part of the scale, with a certain amount of organic matter and silica, according to individual circumstances. Some typical analyses are given in the following table[62]:

TABLE 8
Cooling Water Scale Deposits

	(1)	(2)	(3)	(4)
Calcium carbonate	80·6	48·3	89·2	92·7
Magnesium carbonate	5·3	30·4	—	—
Calcium sulphate	2·0	1·8	3·1	4·2
Iron oxide	4·1	1·5	2·4	1·8
Silica	0·4	1·7	1·2	0·3
Organic matter	7·6	16·3	4·2	1·0

The iron oxide is usually due to rust removed from the metal when the scale is chipped off. As will be observed, the organic matter will often be appreciable; in some cases as much as 30 per cent has been observed.

The scale formation is, of course, due to simple thermal decomposition of bicarbonates of calcium and magnesium under the influence of heat from condensing steam or other source. Since the carbonates have a lower solubility, they are deposited. Such scaling troubles are increased where the cooling water is recirculated through a cooling tower, since the concentration of the salts increases due to the continual evaporation and feeding of make-up water to the system. Scaling is particularly prevalent also where the carbonate hardness of the cooling water is high.

There are several expedients which may be adopted to combat scaling troubles in such systems. The permanent or sulphate hardness usually need not be considered, since scale formation from such salts is unlikely unless a high concentration is reached. The solubility of the permanent hardness salts is very little reduced by the rise in temperature usually experienced. It follows that the make-up to a cooling tower need not be completely softened. Cold lime treatment will be sufficient to precipitate most of the carbonate hardness, and if it is necessary to reduce further still the carbonate content, a treatment with sulphuric acid will convert residual carbonate to sulphate. Alternatively, acid alone can be used, with the advantage of eliminating any need for a softening plant. The acid need only drip into the stream before some system of baffles or other mixing device. In this case it is important to avoid overfeeding acid, otherwise serious corrosion will result. The treatment with acid has the disadvantage that it does not remove any saline matter from the water, so that a higher purge or blow-down from the cooling tower basin may be required. Also acid is much more expensive than lime, and in the absence of any cold lime treatment naturally more acid will be required for a given reduction of carbonate hardness. Another simple method of treatment is to pass the make-up water, or a portion of it, through a zeolite softener.

An economical treatment, extensively adopted in recent years for general industrial cooling waters, is the so-called "Threshold Treatment," the name apparently originating in the very small dosage of chemical required so that the treatment is considered on the threshold of water treatment proper. It has been found that the addition of very small amounts of sodium metaphosphate are effective in delaying the precipitation of calcium carbonate, and also in removing old deposits. As little as 1 to 5 p.p.m. has been found sufficient, in most cases regular operation with less than 2 p.p.m. being satisfactory. In the case of recirculating cooling water systems, the procedure is to limit the concentration of calcium bicarbonate and other salts by blow-down to avoid excessive concentrations. The use of 2 p.p.m. metaphosphate then

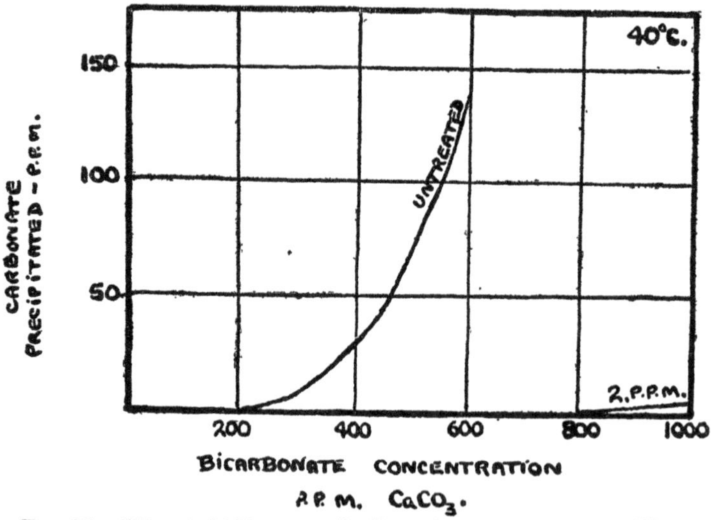

FIG. 17. "Threshold Treatment" effect of Metaphosphate at 40° C.

maintains scale-free cooling surfaces. When the water is used only once, and lower concentrations of calcium are involved, less than 2 p.p.m. is often sufficient. The effect of metaphosphate in preventing deposition is shown in Figs. 17 and 18, the waters being heated for one hour at the temperature quoted. An additional advantage in many cases is substantial reduction in cooling water consumption realized owing to the improved heat transfer conditions in the cooling water system; for example in one case, savings of 40 per cent resulted.

Another method which has often proved successful in holding up precipitation is the addition of organic materials such as glucosides and glucosates. This is probably bound up in some way with colloidal properties.

Scale Formation in Economizers

Suitable water treatment is of particular importance in relation to maintenance of efficiency of the economizer. Apart from the poor heat transfer obtained when pipes are coated with scale, there is a danger of overheating and perhaps failure of one or more pipes. The failure of even a single pipe introduces the risk that the steam generated by the sudden release of volumes of hot water may cause more widespread damage. In at least one case in recent years almost complete destruction of the economizer has been attributed to presence of internal scale on pipes, causing serious overheating and stresses leading to fracture.

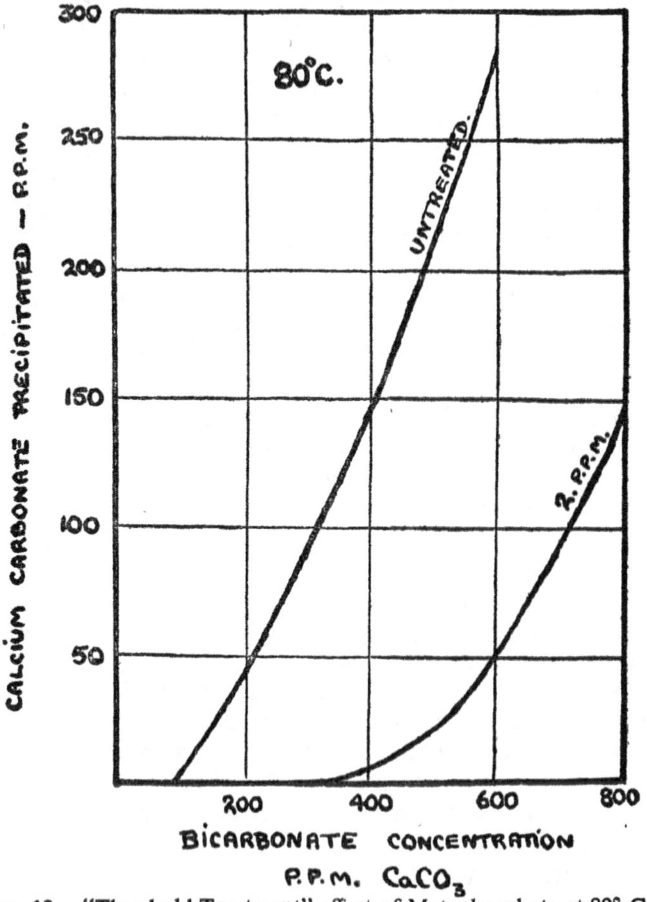

FIG. 18. "Threshold Treatment" effect of Metaphosphate at 80° C.

Scale formation in economizers is primarily due to the temporary hardness in the water, the bicarbonates being decomposed under the influence of the heat extracted from the flue gases. In some cases the economizer is considered as a kind of water purifier as well as an outfit for recovery of heat. Certainly an economizer will deal with much of the temporary hardness in a feed water, but such an attitude is expensive. Most of the decomposition of bicarbonates will occur at the hot end, and such pipes may gradually choke with deposit, leading to poor circulation, and, perhaps, by-passing of some of the pipes. Such conditions may seriously lower the efficiency of the plant. Bulky sediment may also choke some of the lower water boxes. Various factors influence the extent and rate of deposition in the economizer. The quality of the water, particularly as regards the amount of temporary hardness and the relative proportions of calcium and magnesium bicarbonates, is naturally predominant; the temperature rise effected; the rate of flow of the water; the type and efficiency of treatment in use; all these and other factors influence the scaling. In connection with the last-mentioned factor, it will be realized that any "after-precipitation" will lead to fouling of the economizer surfaces, since the heat will facilitate the completion of the delayed softening reactions.

Scaling due to "after-precipitation," residual hardness, and such factors can be combated by the use of sodium aluminate in the lime-soda softening process. As described elsewhere, this reagent has a remarkable effect in speeding up the rate of reactions, and also it effects more complete removal of magnesium due to formation of a less soluble magnesium aluminate. Another advantage is that the presence of aluminate gives greater flocculence to the precipitates, so that they settle more readily, and can therefore be filtered more easily.

Another expedient is to "condition" the softened water with sodium metaphosphate. This is finding increasing application due to a peculiar property it possesses of being able to form complex compounds with calcium. These are soluble, and therefore by feeding metaphosphate in "shots" to give a temporary excess in the water, old calcium deposits can be dissolved. Metaphosphate itself reacts with residual hardness, but not until fairly high temperatures are reached, and therefore serves to condition the water in the boiler. Its property of reacting with alkali first [to produce trisodium phosphate, which then reacts with the residual hardness] also serves to reduce boiler water alkalinity. This is particularly useful, therefore, where boiler alkalinities are rather high. The reagent may be added to the water quite easily as a gravity-fed solution. If some deposition occurs in the economizer due to localized high temperatures, the addition of a little colloid such as cutch (as little as a few p.p.m.) will delay reaction until the boiler is reached.

Electrical Scale Prevention

Numerous attempts have been made over the course of many years to develop electrical methods of water treatment. Such methods are more obvious in combating corrosion, since it is known the corrosion is frequently electro-chemical in nature. Electrical treatments for the prevention of scale formation usually operate on physical principles, attempting to prevent adherence of scale to heating surfaces by changing the condition of the scale-forming matter, repelling the scale particles from the metal, or some similar means.

One form of equipment is a type of "Scale Buoy," which consists of a sealed glass bulb or other vessel containing mercury and a rare gas under several atmospheres pressure. The vessel is supported by brackets in such a way that easy movement is possible in a stream of the water to be treated. Electrical charges are supposed to be produced in the buoy due to the motion, and liberation of these charges from the buoy is claimed in some way to prevent crystalline scale formation, and also to convert existing scale into a loose sludge. Such buoys have been used in boiler plants, and in domestic and hot water heating systems. Cases have been reported where old scale certainly was loosened and removed by water treated in this way, and prevention of scaling has also occurred in some cases. This may be due to scale particles receiving minute electrical charges which prevent coagulation of individual particles due to the repelling effect of similar electrical charges. Building up of scale would then be prevented.

Another method is to introduce a small current into the boiler shell. Electrolytic effects and possible attendant chemical reactions should be neutralized by this external electromotive force, and it is claimed that after proper investigation of each individual case corrosion will be prevented, and scale will not deposit on the heating surfaces; instead, scale-forming matter can be removed with the boiler blow-down. The efficiency of the system is said to be governed by careful elimination of short-circuiting effects of design features on the boiler, a study of the behaviour of the water composition (which, of course, varies from case to case), the selection of the best positions on the boiler for introducing the neutralizing current, and the choice of the correct current for individual conditions.

The Pulso process also produces an electrical charge on the boiler shell. In this system a special type of impulsive charge is produced by an electronic valve equipment. The pulses of bi-polar static charge so produced on the shell are claimed to prevent scale formation, since it destroys the natural attractive forces between the boiler shell and particles of scale-forming matter. According to this, scale is formed by the flocculation of colloidal particles of scale-forming matter, contact with the boiler shell serving to neutralize the minute electrical charges on the colloidal particles, which therefore flocculate or build up as scale on the metal. The bi-polar charge on the shell is said to repel the particles, which thus have to remain as a sludge in the body of the boiler water. The Pulso valve equipment is said to give the best type of charge to the shell to achieve these objects, a number of contacts being necessary to obtain a general, rather than a localized, effect.

The O.C.P. process operates on a different principle. This outfit consists of specially constructed electrodes housed in a container through which the water flows. A minute alternating current is applied to the water as it flows in fine streams closely over the electrode surfaces at a predetermined rate. The current is said to change the crystalline nature of the saline impurities so that they lose their scale-forming properties. On production of steam, the saline matter remains in the boiler water as a sludge which may be removed by blow-down. The theoretical explanation of the action is as yet not fully known, but cases have been reported where old scale deposits have been removed, in addition to absence of formation of new scale. Finally, it may be mentioned that current requirements are very small, and may be met by taking off either power or lighting circuits.

Removal of Scales and Deposits

Periodical removal of scale, sludge, slime, and grease deposits from condensers, economizers, and boilers is one of the major problems of plant operation. Such deposition occurs even in the best regulated plants at times due to such factors as residual hardness due to solubility limitations of softening reactions, imperfect softening as a result of variations in raw water quality or lapses in softener control, etc. If operating efficiency is to be maintained, the deposition must be removed either by treatment with chemical solvents or by the use of mechanical equipment.

Mechanical Methods.—The oldest method, used particularly for cleaning tubes on condensers and heat exchangers, is to shoot rubber plugs through the tubes from a specially designed gun operating with an air or water pressure of 50–100 lb. The simple rubber plugs are cheap and can be used repeatedly, but are effective only for dealing with sludge, slime, and loose scale. A development of this idea is a double-fluted plug with a bronze scraper attachment which is used with a pressure of 100–150 lb. It takes about three times as long to clean tubes with such scraper plugs, but they are effective in cutting away hard scale coatings. For dealing with very hard deposits, a variety of special scaling outfits are available. These comprise sets of scaling heads in the form of drills, reamers, abrasive wheels, and brushes, which can be operated inside the tubes through a flexible drive from a small $\frac{1}{2}$ h.p. motor (electrical or compressed air) or a small water-driven turbine. While the water-driven equipment has low initial cost and maintenance charges, much depends upon the hardness of the scale and the pressure of water available to drive the turbine. It is usually more suitable for the softer types of scale. An advantage is that the scale can be washed away by the water as removed; where the water is circulated over and over again via a pump, however, suitable strainers are required to avoid choking the latter. Compressed-air-driven equipment has the disadvantage, more notable in enclosed spaces, that scale dust may be blown about, but the method is very efficient and convenient since a little portable compressor can be moved about the plant very easily. Electrically-driven outfits too have the advantage of mobility, and are also quite clean and conveniently used from wall plugs, lamp sockets, etc. A jet of water can usefully be provided to cool the equipment and at the same time wash away the scale as removed. In some cases condenser tubes are still cleaned by sandblasting with air, but a common difficulty, if tubes are at all long, is that the centre of the tube is not properly cleaned even if the blasting is done from each end.

Chemical Methods.—Chemical removal of scale operates by solution of the deposit in acid, and in the past risk of metal corrosion has militated against adoption of this procedure. Many of the past failures, however, have been due to lack of careful study of the character of the deposit to be removed and to use of unsuitable solvents. In recent years chemical cleaning has been widely adopted for all classes of power plant equipment, owing to the development of effective inhibitors for limiting metal corrosion without materially reducing the solvent action on scales. These range from the common glue and copper sulphate of early solvents to a host of organic materials, such as anthracene, pyridine, thiourea, quinoline, dextrine, aldehydes, mercaptans, etc., which have been patented as modern ingredients. Intensive research over the last decade indicates that the best inhibitor varies according to the nature of the scale, the metals in the plant to be cleaned, the temperature of treatment, and similar practical factors. In addition, it has been found useful to include in the mixture other reagents—wetting agents, catalysts, reaction-control agents, intensifiers, etc.—to give the acid solution the optimum effect on the particular type of deposit to be removed. This intensive study of chemical descaling has resulted also in the development of an elaborate technique of preliminary tests to ensure that the chemicals selected are those best suited to the individual plant conditions.[63, 64]

The advantages of chemical solution of deposits are: (*a*) more complete removal of deposition, even in parts inaccessible to abrasive tools, owing to ease of fluid penetration; (*b*) avoidance of dismantling of equipment since the solvent penetrates wherever steam and water flow—this greatly reduces labour costs and also saves material, such as new gaskets; (*c*) plant outage time is substantially reduced; (*d*) the metal surface is left clean of old scale nuclei which would assist new scale deposition; (*e*) the metal loss is negligible—in at least one instance this has been confirmed[65] by an actual measurement of the iron and copper content of scale removed from condenser tubes by acid and by turbining, the amount of metal removed from the tubes being the same by each method.

The usual solvent is hydrochloric (muriatic) acid, the concentrated reagent being diluted with water to about 2 per cent strength by volume and suitably inhibited, etc. At one time the maximum temperature considered advisable for the treatment was 150° F., but with modern improved inhibitors it is found safe to operate up to 180° F., which results in more rapid reaction. The usual precautions observed in handling acids should be taken, such as use of rubber gloves and goggles, with protective rubber aprons; it is also useful to have respirators available in case they are needed in dealing with the hot solution. Acid should always be poured slowly into water, and one should never pour water into acid, when preparing dilute solutions. An alternative which has been used occasionally is an organic acid such as lactic acid,[66] again made up to about 2 per cent strength by volume, but with this reagent an inhibitor is not necessary. Another solvent, recommended by the Russians,[67] is chromic acid, which also does not require an inhibitor, since it has a negligible effect on iron. It requires a longer time of treatment (2–3 times) and a higher temperature (200° F.) as well as 4 per cent strength (on basis of CrO_3) for satisfactory solvent action. It is also a more expensive reagent and requires rather more care in handling, since it attacks the skin, mucous membrane, etc.

Treatment procedure should be varied to suit plant conditions such as lay-out, character of scale, etc., and careful preliminary tests and planning are advisable to enable the most suitable technique to be worked out. The most important preliminary is to determine the exact chemical and physical characteristics of the deposition, and very careful sampling is essential if the results of these tests are to be accurate as a

TABLE 9

Types of Scale Deposit

Name of scale	X-ray analysis	Source
Calcium-hydroxy-phosphate	$Ca_{10}(OH)_2(PO_4)_6$	Boilers, Economizer.
Analcite	$NaAlSi_2O_6.H_2O$	Water tube boiler.
Pectolite	$4CaO.Na_2O.6SiO_2.H_2O$	Condenser.
Serpentine	$H_4Mg_3Si_2O_9$	Water tube boiler.
Iron oxide	Fe_3O_4	Oil cooler.
Hydrated magnesium silicate	$Mg_3Si_2O_7.2H_2O$	Fire-tube boiler.
Vermiculite	$H_3Mg_3Al(SiO_4)_3$	Water tube boiler.
Noselite	$Na_8Al_6Si_6O_{24}SO_4$	Heat exchanger.
Aragonite	$CaCO_3$	Evaporator, Feed heater.
Anydrite	$CaSO_4$	Heat Exchanger, Boiler.
Haematite	Fe_2O_3	Water tube boiler.
Magnesium hydroxide	$Mg(OH)_2$	Evaporator.
Magnesium oxide	MgO	Boiler.

guide. Ideally, the sample should represent the most dense scale, such as found at the hottest parts of the plant, and must be representative of the full thickness of the scale as it has deposited on the metal. If the deposit proves to be a simple compound, readily soluble, a suitable solvent can be selected without delay, giving due consideration, however, to the various metals which will be exposed to the acid. When the more difficult scales are encountered, modern methods are to carry out a full chemical analysis. These laboratory tests give full information on the deposit's porosity, density, extent of lamination, ease of disintegration, chemical composition, and of course solubility at various temperatures in various solvents. In addition, an X-ray analysis is done to determine the crystal structure and form of chemical combination, followed by a spectrographic analysis if necessary to check the metallic elements present. On the basis of this detailed knowledge, the optimum strength and type of solvent can be selected. The development of such careful detailed tests has been enforced by the wide variety in types of deposits experienced in modern plant (see Table 9) and is also essential to safeguard expensive equipment against damage which might result from faulty treatments based on inadequate information.

General experience dictates the use of weaker solutions for a longer time rather than stronger solutions for a shorter time. Solutions should be heated by a fire under the boilers, steam injector, or other means, and definite circulation by pumps gives a useful scouring effect which accelerates the cleaning. Plant can be treated in sections if convenient, using rubber plugs to seal off parts not to be treated. Usually, descaling is completed in 4–8 hours, though up to 18 hours may be required, according to plant conditions; in most cases plant can be put back into service in 24 hours. The strength of the cleaning solution is determined on samples taken periodically, and treatment ends when this remains constant, indicating that reaction is completed. Finally, the equipment must be cleared of acid by flushing first with water, then with weak soda ash solution to kill last traces of acid, and finally with water again.

Chemical cleaning has been used in a wide range of industries with complete success on all types of heat exchange equipment, as well as power plant. As examples of the latter may be quoted a 425 lb. water-tube boiler cleaned in 6 hours with inhibited hydrochloric acid,[68] B. and W. 365 lb. pressure boilers cleaned in 18 hours at Kearney power station,[69] a 7,280-tube condenser cleared of calcium carbonate in 4 hours,[65] two 350 h.p. turbines in which scale had reduced the output to less than 50 per cent of the rating,[70] a 1,250 lb. central station boiler completely cleaned of a deposit of calcium-hydroxy-phosphate and iron oxide and an economizer fouled with a similar deposit which was removed in less than 8 hours after unsatisfactory attempts at mechanical cleaning.[63]

Grease Removal.—Slime and grease deposits in condensers can usually be dealt with by a weak solution of caustic soda (normally 2 oz. to the gallon, but up to 7 per cent solution if deposition is bad). Sometimes ½ oz. of cyanide per gallon is added to accelerate cleaning, and proprietory mixtures often contain wetting agents, phosphates, and metasilicates which greatly improve the cleaning. Circulation and heating (not above 150° F.) is desirable.

Miscellaneous Treatments.—Kerosene oil is often used for removal of existing scale deposits, and more occasionally for preventing the formation of fresh deposition, but opinions differ considerably regarding the desirability of the practice. The action of the kerosene is probably more mechanical than as a solvent. The greater penetrating powers of kerosene as compared with water are well known, and the writer's opinion is that the oil creeps behind the scale and "cuts" it away from the metal in much the same way that it "cuts" the rust between seized metal surfaces. Chemical effects are very uncertain, since the combined influence of high temperature and pressure, with caustic alkalinity in the boiler water, are unknown. Paraffin hydrocarbons in the oil can be converted to fatty acids and their sodium salts, and apart from the corrosion aspect it is known that fatty organic acids tend to promote priming and production of wet steam. The extent of such a conversion, if any, will depend upon boiler conditions, but certainly high temperatures and pressures will favour its development. At high temperature, too, there will be a greater risk of distillation of the kerosene with the steam, and joint trouble has been experienced due to gaskets being attacked by kerosene so carried over. It is evident from these considerations that until experience has been gained at least under individual plant conditions it should be used only very cautiously—not at all at high pressures, and only periodically at low pressures.

REFERENCES

(1) Partridge & White, " Ind. Eng. Chem.," 1929, p. 839.
(2) Anderson, " The Engineer," Nov. 1927, p. 605.
(3) C. E. Joos, " Combustion," Sept. 1942, p. 45.
(4) Joos & Kemmer, " Combustion," April 1943, p. 45.
(5) " Ind. Eng. Chem.," May 1940.
(6) " Engineer," Aug. 1940, p. 79.
(7) British Patent 555,535.
(8) Christman, Holmes & Thompson, " Ind. Eng. Chem.," 1931, p. 637.
(9) C. C. Spilman, " Power," Jan. 1929.
(10) Beal & Stevens, " Chem. & Ind.," 1931, p. 307T.
(11) Clark & Price, " Chem. & Ind.," 1933, p. 35T.
(12) Adams & Holmes, British Patent 450,308/1934.
(13) Myers & Eastes, " Ind. Eng. Chem.," 1941, *33*, p. 1203.
(14) Gilwood & Calise, " Power," June 1945, p. 107.
(15) Tiger & Soussman, " Ind. Eng. Chem.," 1943, p. 187; Tiger, " Trans. Am. Soc. Mech. Eng.," 1938, p. 315; S. J. Broderick, Bureau of Mines Report No. 3571 (June 1941) covers the subject comprehensively.
(16) D. K. French, " Ind. Eng. Chem.," 1934, p. 1319; " Boiler House Review," March 1935.
(17) Hall, " Power," Feb. 1926.
(18) Hall, " Ind. Eng. Chem.," 1929, p. 826.
(19) Hall and collaborators, Bulletin No. 24, Mining & Metallurgical Investigations, Carnegie Institute of Technology, Pittsburgh.
(20) Hall, " Ind. Eng. Chem.," 1925, p. 283.
(21) Hall, " Trans. Inst. Amer. Chem. Eng.," 1924.
(22) Hall, Nat. Elect. Light Association Serial Report No. 278–81, p. 46 (1928).
(23) Clark & Hunter, " J. Chem. Soc.," 1935, p. 383; F. G. Straub, " Ind. Eng. Chem.," 1932, pp. 915, 1175.
(24) Noyes, Carnegie Inst., Washington, Pub. 63, p. 1907.
(25) Splittgerber, " Die Wärme," 1928, p. 733.
(26) Clark & Gerrard, " J. Soc. Chem. Ind.," 1938, p. 295.
(27) Anderson, J. Inst. Fuel, 1928, p. 131.
(28) " Ind. Eng. Chem.," Jan. 1939, p. 58–63.
(29) Gilmore, " Ind. Eng. Chem.," 1937, p. 584.
(30) Wirth & Joos, " Combustion," Dec. 1944, p. 45.
(31) " Die Wärme," 1934, pp. 630, 644; " Engineering & Boiler House Review," March 1937.
(32) Partridge & Purdy, " Ind. Eng. Chem.," Vol. 30, p. 387.
(33) Jander & Heukeshoven, Z. Anorg. Allgem. Chem., 1931, p. 361–82.
(34) Stumper, " Die Wärme," 1932, p. 272–8.
(35) Lindsay & Ryznar, " Ind. Eng. Chem.," 1939, p. 859.
(36) Schwartz., J. Am. Waterworks Assoc., 1938, p. 551.
(37) See reference 8.
(38) Straub, " Ind. Eng. Chem.," 1936, p. 36.
(39) Geisler, " Vom Wasser," Jahrbuch Vol. 12, 1937.
(40) Betz., " Combustion," Dec. 1940, p. 28; Tiger, ibid, p. 27.
(41) British Patent 542,751; Gerrard, " Power & Works Eng.," Oct. 1943.
(42) S. T. Powell, " Paper Trade Journal," 1938, Sect. 89.
(43) Haering, " Power Plant Eng.," 1938, p. 331.

(44) S. E. Tray, " Mech. Engineering," 1938, p. 475.
(45) F. G. Straub, Bulletin No. 282, Univ. Illinois Eng. Exp. Station, July 1936.
(46) C. H. Fellows, J. Inst. Fuel, 1941, p. 85.
(47) H. Goerke, " Elektrizitaetswirsch," 1939, p. 614.
(48) D. W. Rudorff, " Cheap Steam," May 1943, p. 78.
(49) Splittgerber, " Archiv. für Wärmewirtschaft." March 1941.
(50) Straub & Grabowski, Trans. Am. Soc. Mech. Eng., 1945, p. 309.
(51) F. Spillner, " Chem. Fabrik.," 1940, p. 405.
(52) Webb & Call, " Combustion," June 1943.
(53) Hall, abstract Trans. Am. Soc. Mech. Eng. paper in " Power," Dec. 1943.
(54) Hall, Trans. Am. Soc. Mech. Eng., July 1944, p. 457–489.
(55) Webb, Parkes, Trans. Am. Soc. Eng., July 1945; Baker & Hankinson, " Combustion," Dec. 1944.
(56) " Cheap Steam," July 1932.
(57) " Engineering & Boiler House Rev.," Dec. 1935, March 1938.
(58) Waddington & Clark, in a paper read before the West of England Division of the Papermakers' Association of Gt. Britain on Feb. 22, 1935.
(59) W. Welch, " Power," Nov. 1935, p. 611.
(60) R. B. Martin, Trans. Am. Soc. Mech. Eng., 1938, p. 475.
(61) " Power Plant Eng.," May 1941.
(62) Matthews, " Cheap Steam," Jan. 1939.
(63) Alquist, Groom & Williams, Trans. Am. Soc. Mech. Eng., 1943, p. 719.
(64) M. M. de Witt, " Combustion," Oct. 1943, p. 41.
(65) M. Gelfand, " Power Plant Eng.," Oct. 1941, p. 64.
(66) Drabelle, " Combustion," April 1941, p. 47; " Power," June 1941.
(67) " Steam Eng.," Aug. 1938.
(68) M. E. Brines, " Power Plant Eng.," Vol. 49, p. 90.
(69) " Electrical World," Vol. 118, p. 95.
(70) Entrikin, " Ind. Eng. Chem.," 1938, p. 1279.

PART THREE

CORROSION

THE GENERAL SUBJECT OF CORROSION HAS RECEIVED CONSIDERABLE attention from chemists and engineers for many years and innumerable research programmes, many of them elaborate and extending over years, have been carried out in an effort to conquer the trouble. So many factors influence corrosion, however, that progress has been slow and only in the last twenty years has a reasonably satisfactory answer been found for the extraordinary manner in which corrosive attack often makes itself evident. For example, none of the theories put forward in the last century was able to explain the local and apparently capricious nature of corrosion, which is exemplified in boilers as pitting. It is not proposed in this book to discuss all the factors involved. or even most of them, for they are many, and abler writers than the author have written volumes on the subject. It is hoped rather to deal briefly and practically with such factors as play a predominant part in corrosion of steam power plant.

The serious nature of corrosion in boiler plant will be admitted by everyone and most power plant operators will agree that its prevention is an even more urgent problem than that of scale prevention. Formation of scale on heating surfaces soon makes itself evident, but corrosive attack is frequently unnoticed until considerable damage has been done. It may be emphasized at this time, too, that prevention of scale does not always mean absence of corrosion ; plates and tubes may be clean merely because they are being attacked by acid which dissolves any scale. Clean plates and tubes can, however, be maintained free from corrosion and this is accomplished by careful attention to the composition of the feed water.

Early Theories of Corrosion

At first sight the whole thing seems extremely simple. It was observed that iron rusted and that the rust was oxide of iron. It was only natural, therefore, that the first explanation to be put forward was that corrosion is merely straightforward oxidation. This was overthrown by the discovery that moisture was necessary also, since iron will not form rust in a perfectly dry atmosphere. Since this earliest idea various explanations have been developed. The *acid theory of corrosion* stated that the presence of carbon dioxide or some other weakly acid substance was needed for the rusting of iron. It was considered that the carbon dioxide reacted with the iron to form soluble ferrous carbonate, which was then oxidized to ferric hydroxide (hydrated rust), the carbon dioxide being again set free to attack a further quantity of iron. Similar cycles were said to occur with other weak acids. This certainly occurs in the case of carbon dioxide, as is shown later, but that corrosion in general is due to such action is not true. The idea was eventually overthrown as a general explanation because it was found possible for iron to rust under conditions in which carbon dioxide could not possibly be present, and also it was found that faintly alkaline solutions of certain salts would corrode iron and steel. Another line of thought resulted in the *hydrogen peroxide theory*, which suggested that the first stage of corrosion was the production of hydrogen peroxide, this being the real corrosive agent. The idea originated because it was found that hydrogen peroxide was formed in the wet oxidation of many metals, and the fact that chromates and alkalis inhibited attack was said to be due to the destructive action of these substances on hydrogen peroxide. The idea was effectively put out of court by the work of Moody. According to the theory, hydrogen peroxide should naturally have the most corrosive effect upon iron, whereas Moody showed that this was not the case with neutral peroxide. Friend's *colloidal theory* is a recent development, and according to this the rusting of iron is catalytically promoted by colloidal iron hydroxide, which acts as an oxygen carrier. This does not account for the fact that certain colloids retard corrosion, whereas actually they should catalyse the action in the same way as the ferric hydroxide is supposed to do. The *electrochemical theory* is another explanation. According to this iron has a natural tendency to pass into solution in water but it will only actually dissolve if there is present a second substance with less "solution tension," or lower tendency to pass into solution, than iron itself. Thus if the solution is sufficiently acid, hydrogen will come streaming off the metal and corrosion will be rapid, i.e., the iron will dissolve or pass into solution. If, however, the solution is not sufficiently acid, the hydrogen will only slowly diffuse away *in solution*, so that corrosion will be much slower. The presence of oxygen in solution assisted corrosion by acting as a depolarizer and removing hydrogen, thereby speeding up corrosion.

None of these theories explain rationally the problem of localized attack, such as pitting or why corrosion should proceed more rapidly at the bottom of a deep pit than elsewhere and so penetrate the metal before there is any appreciable corrosion elsewhere. They do not explain why the products of corrosion sometimes form a compact layer on the metal and stop any further attack, whereas at other times a porous or flocculent corrosion product sometimes produced stimulates the attack.

Corrosion by Differential Aeration

All the above theories of corrosion have now been superseded by the electrochemical theory, depending on the principle of differential aeration. This is well fitted to meet the extremely diverse situations that are encountered to-day. This statement will be admitted when a few of the cases already explained are mentioned. The theory has been found capable of explaining numerous cases of intense localized corrosion, and also of interpreting the action of drops of liquids on metals. It has been adopted to explain the corrosion of condenser tubes, and also the attack of galvanized iron sheets by sea spray when stacked on board ship. Finally, the principles of the theory have served to account for the corrosion of certain parts of a coal gas holder. These few cases are sufficiently diverse to satisfy the most exacting critic so that this modern conception of corrosion may be accepted with every confidence.

The principle of differential aeration has been put forward by U. R. Evans in recent years, but it was believed by many chemists early in the last century that corrosion was an electrical phenomenon. The old electrochemical theory has now been combined with the principle of differential aeration, and it may be stated at once that the more destructive troubles of the modern engineer are due to this differential aeration. This differential aeration was enunciated by Evans as a result of what he terms the "key experiment." In this experiment two clean strips of iron cut from the same sheet were placed in a cell containing a solution of salt, the cell being divided into two compartments by a porous partition. Each compartment contained one strip, and the two strips were joined, one to each terminal of a milliammeter. Then, although each strip was of the same composition, and was exposed to the same solution, when air was bubbled around one strip an electric current was set up. This was indicated on the milliammeter, and the aerated electrode was the negative pole or cathode. Also, the unaerated electrode, or anode, lost weight, i.e., it was corroded. The dissolved salts, of course, ionize and carry the current generated. Evans has shown that such differential aeration currents can be set up not only with iron electrodes, but also with electrodes of zinc, cadmium, lead, nickel, or aluminium. The principle therefore has general application.

Dissolved Oxygen

The interpretation of this experiment with reference to the role of dissolved oxygen in corrosion is obvious. If, from any cause, one portion of a plate of metal immersed in a solution becomes wholly or partially inaccessible to oxygen (whether dissolved in the solution or present in the atmosphere), then that part of the metal becomes the anode, or attackable pole, of a cell, while the aerated portion of the metal serves as the cathode. Active corrosion will then occur at the anodic area. The apparent contradiction that oxygen should cause corrosion by *not* reaching parts of the metal is explained by the fact that with perfect distribution of oxygen a protective oxide film would be formed to stop the attack. The attack which occurs is due to the lack of this film and subsequent establishment of small electric cells as explained. It will be interesting to consider some common types of corrosion and their explanation from this standpoint.

Consider first the behaviour known as pitting. The attack in such a case begins at pores and crevices in the metal, which are specially inaccessible to the metal. The interior of such a crevice thus becomes anodic, and is attacked. At this stage two things may happen. The rust formed from the pits may fall and blanket the metal below, which is thus screened from oxygen. The corrosion will then spread over a wider area and so become less dangerous. If, however, conditions are such that the rust produced by the corrosion remains near the corroded part, then

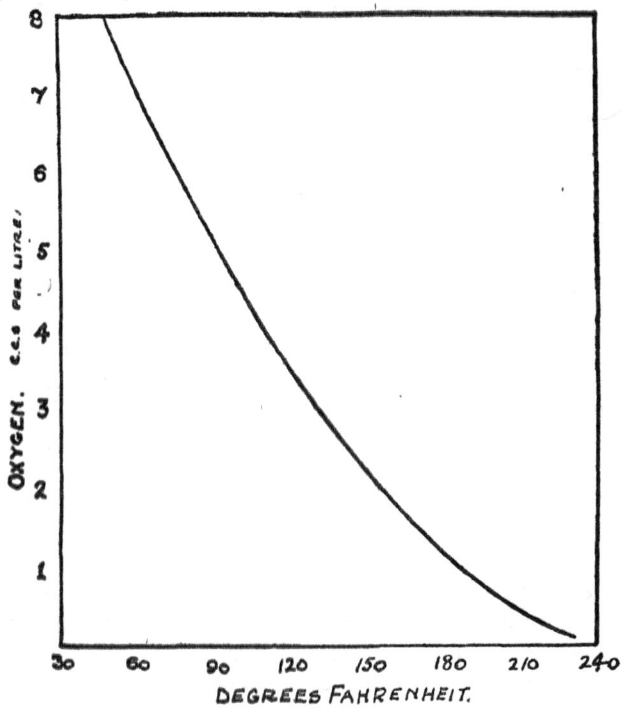

FIG. 19. Solubility of Oxygen in water at various temperatures

it will serve to screen still further that portion of the plate from oxygen, and even more intense corrosion will be promoted. Hence the pitting which occurs.

The great objection to the formation of a thin protective scale as a means of preventing corrosion is that it cannot be controlled either as regards thickness or location. Differential aeration conditions will obtain if part of the boiler heating surface is covered with scale and another is not. Even supposing that all the metal were covered, then as soon as the scale becomes cracked, aeration will be possible there and pitting is likely to occur beneath the scale. Apart altogether from the loss of fuel and liability to burst tubes, the formation of a "thin protective scale" is not likely to afford any real lasting protection.

There are always periods when plant is lying idle and at such times a steel vessel will contain a stagnant water-line. Even if the vessel has been emptied there will always be pools of water left at the bottom, and if left filled the vessel is sure to contain some air trap at the top. Water line conditions are therefore well worth consideration. It is commonly noticed that a sheet of metal immersed in stagnant water is corroded below the water line. The principle of differential aeration explains why it is usual for the corroded area to begin some little distance below the water-line. Immunity from attack at the water-line and just below it is due to the fact that that portion of the metal is accessible to oxygen and becomes covered with the usual protective oxide film. On the other hand, lower down the metal is less accessible to oxygen and differential aeration obtains. In this region the attack begins at pores and crevices particularly inaccessible to oxygen, and according to conditions as outlined above, the corrosion becomes either general and non-severe or localized and intense.

Finally, there is the situation when the water is moving. It has been shown that under these conditions corrosion tends to become localized. This is natural, since there is better oxygen distribution and it is only the more inaccessible spots which become anodic and therefore corroded.

The various methods of reducing corrosion will now be evident. Salts may be added to the water to form protective films on the metal or to otherwise inhibit corrosion, the oxygen may be removed from the water either chemically or by physical means, or electrical methods may be adopted to prevent any part of the boiler metal from becoming anodic.

The treatment of water with various chemicals which are supposed to inhibit corrosion is widespread. Most engineers are familiar with the idea that chromates and dichromates are good inhibitors, and one or other of these minerals has often been adopted in cases of severe corrosion. These compounds are oxidizing agents and they protect the metal by coating it with a protective oxide film. The application of inhibitors requires great care, however, since individual plant conditions may vitally affect the success of the treatment. Thus, several investigators have found that in the presence of chloride the use of potassium dichromate is inadvisable owing to risk of severe local corrosion being set up.

De-Aeration by Chemical Methods

All chemical methods are based upon the same principle ; some easily oxidized substance is used to take up the oxygen.

A widely-used material is iron. While it may be used in the form of sheets and plates, a greater surface area for absorption of the oxygen, and consequently more efficient de-aeration, is available when the metal is in the form of wool or shavings. Another point is that pure iron does not rust as quickly as impure samples, and it is therefore advantageous to use the latter for quick removal of oxygen. It has been found that the presence of manganese in iron accelerates the rate of rusting or absorption of oxygen, the best material containing rather less than 1 per

PLATE 12. PERMUTIT SODIUM AND HYDROGEN ION SOFTENER

PLATE 13. B.B.A. DEJECTOR PLANT
(Six units in a Yorkshire textile mill)

PLATE 14. CHLORINATING PLANT FOR INTERMITTENT DE-SLIMING OF CONDENSER TUBES
(*Boiler House Review*)

cent. Graphitic carbon, on the other hand, reduces the activity of the iron. A plant which uses iron wool is the Kestner degassing plant, and several factors affect the output of such a unit. The effect of surface area has been pointed out, and this is taken care of in this case by using the iron in the form of wool. There is also another point in connection with the speed of reaction, and that is that the absorption does not proceed at the same speed from start to completion. The first portions of the gas are absorbed much more quickly than in the latter stages of absorption; this is due to a protective coating formed on the iron which leads to poor contact between metal and gas. After treatment this protected iron becomes available again, so in the Kestner plant only one half the steel wool is used at one time, while the other half is being regenerated. The water flow is reversed daily and regular removal of the gas results. A filter is fitted to prevent any rust passing to the boilers. The upkeep of the plant is low, since the consumption of steel wool is of the order of an ounce per thousand gallons of water treated. Two other factors in the absorption which may be mentioned are the temperature of the water and the rate of throughput. The higher the temperature the more rapid the absorption; consequently it is usual to heat the water before passing it through the degasser. As regards throughput, it will be obvious that the feed can be passed through too rapidly to take up all the oxygen and the velocity of the water must be adjusted to suit the size of plant installed. A low throughput and satisfactory oxygen removal is decidedly better than a large throughput and a "risky" amount of oxygen.

Colloidal organic materials have also been used to combat dissolved oxygen, either by incorporating materials which will absorb the oxygen chemically, or by covering the metal surface with a thin protective film (such as the black film of iron tannate sometimes deposited from alkaline tannin solutions) claimed to resist access of dissolved gases to the metal. Colloidal tannin extracts, for example, are chemically similar to the pyrogallol used in flue gas analysis for oxygen absorption, and in fact in a comparison of pyrogallic acid with some tannin extracts it was found that absorption varied from a minimum of 17 per cent for liquid spruce to 80 per cent for chestnut oak extract. This was at ordinary temperature and pressure; increases in alkalinity, temperature, and pressure substantially increase and hasten the absorption. This oxygen absorption works very well in practice, as illustrated by a plant operating at 1,200 lb. pressure,[1] which was given treatment with a paste made up of 200 lb. chestnut oak extract and 250–300 lb. of caustic soda and disodium phosphate (proportions 40:60) in a 600 lb. barrel. Daily tests showed complete stoppage of corrosion. There is also some evidence that organic colloids reduce the ionization of minerals such as sodium chloride, and therefore combat electrolytic corrosion.

Another reagent for the removal of oxygen which has been described recently is sulphur dioxide, which may be added as the free acid or as sodium sulphite or bisulphite. These substances are reducing agents and

G

consequently absorb the oxygen and so prevent corrosion. An important advantage of sulphite absorption is that a stable substance, sodium sulphate, is formed by the reaction with oxygen, whereas organic materials such as tannins decompose to produce carbon dioxide in the boilers. At lower temperatures, and in dilute solutions, the sulphite reacts only slowly with the dissolved oxygen and as a result unreacted oxygen may be present for some time, especially in feed-lines, economizer, etc. One way of dealing with this is to give the water at least twice the equivalent dose of sulphite to assist rapid reaction; another method is to add 1 or 2 p.p.m. of catalysts such as nickel sulphate, which will speed up the reaction.

The use of ferrous sulphate is increasing, particularly in cases of high boiler alkalinity. This chemical reacts with alkalinity to give a precipitate of ferric hydroxide and three things are therefore accomplished at the same time, viz., the alkalinity is reduced, coagulation is effected, and oxygen is removed owing to the powerful affinity of the precipitate for the gas. A similar expedient is the use of ferrous hydroxide suspension which is recommended[2] for the complete removal of dissolved oxygen.

De-Aeration by Physical Methods

Agitating and spraying of water is a physical device but it will ensure only that the water is not supersaturated. Dissolved oxygen may also be removed in a properly operated ordinary open or mixing vented feed water heater. The efficiency of the removal depends upon the size of the heater and the manner in which the heater is vented. If the water is kept at or near the boiling point in a well-vented heater, the results will be satisfactory for the ordinary boiler plant. The vents should be so arranged that accumulations of released gas in pockets in parts of the heater will not occur. It is advisable also that the vents should rise as nearly vertical as possible, since condensate may collect in any horizontal portions or parts which slope even slightly downwards. Such condensate may seal the vent, and thus prevent escape of the liberated gas. For the same reason throttling valves on the vents should be regularly examined to be sure they do not prevent escape of the gas. Such an outfit is supplied by Messrs. William Boby and by Messrs. Permutit Co., and the design usually incorporates arrangements for maintaining the supply of steam sufficient to keep the water at about the boiling point.

In other physical plants, application is made of the fact that dissolved gases are most easily liberated from finely divided water at high temperatures. Thus, in the "flash" type of de-aerator, the previously heated water is sprayed into an evacuated vessel and the degree of evacuation and amount of pre-heating are so arranged that the water boils on entering the vessel, and so liberates the gas. This operation can be accomplished in one or more stages. Thus, Weir's have a two-stage plant in which the water first passes into a chamber at a temperature about

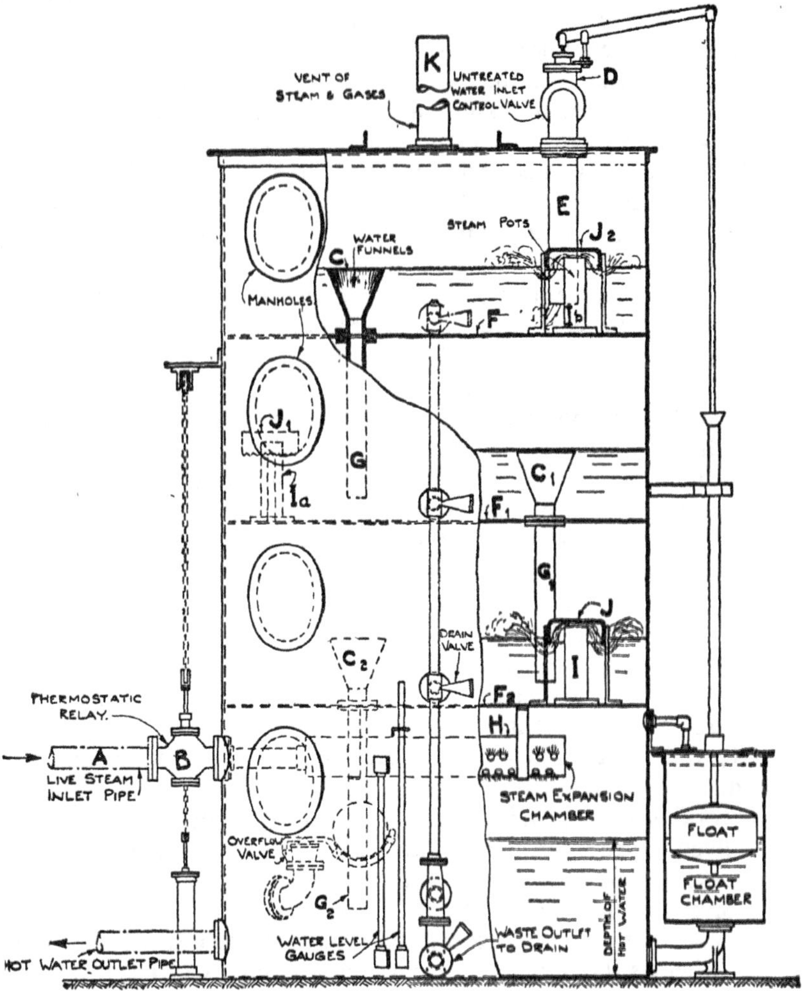

FIG. 20. General arrangement of Boby De-aerator.

The water funnels C, C_1, C_2, maintain a certain depth of water in each of the compartments formed by the partitions F, F_1, F_2. During admission of untreated water via the control valve D, water passes over the edges of the funnels to the base of each compartment by means of the dip pipes G, G_1 and G_2. Thermostatically-controlled live steam enters at A in amount sufficient to maintain a water temperature of 212° F. The steam enters via the expansion chamber H and passes through the riser I into the inverted pot J. The resultant blowing of steam through the water causes agitation, and in addition this action in all parts of the outfit produces boiling throughout the surface water, thus releasing the dissolved gases.

5° F. below that of the feed. This results in a liberation of part of the dissolved gas. The released air and vapour from this first chamber serve to disintegrate the water in a second chamber and almost all the dissolved gas is then liberated, and the water is then withdrawn and pumped to the boiler. In the Neckar de-aerator the water is sprayed into a partially evacuated vessel via the atomiser, and the released gas removed by an ejector. This plant is interesting in that the ejector is operated by steam obtained from an expansion chamber placed in the Neckar continuous blow-down boiler water return line. This eliminates the need for any extraction pump which may give trouble due to leakage back of air.

A new type of de-aerator has been described for handling treated water which may be corrosive, such as sodium-zeolite-softened effluent which has been treated with acid, or effluent from hydrogen zeolite plants which contain acids and considerable amounts of dissolved carbon dioxide. In this method, instead of passing the water over trays which might be corroded, it is atomized by a high-velocity jet of steam like oil in an oil burner; the water is first pre-heated by spraying it through a steam atmosphere, after which atomization with steam blasts breaks it up into a very fine state of sub-division which allows release of the dissolved gases. Data published[3] for this method of de-aeration show it to be very effective in dealing with high initial quantities of dissolved gases. Appreciable conversion of bicarbonates to carbonates is also obtained, figures being given showing that the pH value of the water increased from approximately 5·7 to 8·7; at the same time an oxygen content of about 8 ml. per litre was reduced to less than 0·01, and the carbon dioxide (free) content of 30–35 p.p.m. was eliminated.

Plants working on a physical basis do not give water completely free from dissolved oxygen, but the amount of gas left in the feed is claimed by most makers to be less than 0·1 ml. per litre. This may be considered non-corrosive for normal plants. The flash type de-aerator has the advantage over the chemical absorption type that the cost of upkeep is negligible, whereas the latter requires a constant supply of chemicals.

Corrosion due to Unstable Salts

Corrosion troubles may also be caused by certain minerals commonly found in natural water supplies, even though they may be present only in small amounts. This is due to the fact that they are unstable under boiler conditions and break down to give free acids. Magnesium salts may be particularly troublesome in this respect.

Magnesium chloride can break down, or hydrolyse, as it is termed by chemists, according to the reaction

$$\underset{\substack{\text{Magnesium}\\\text{chloride}}}{MgCl_2} + \underset{\text{Water}}{2\,H_2O} \rightarrow \underset{\substack{\text{Magnesium}\\\text{hydroxide}\\\text{(deposit)}}}{Mg(OH)_2} + \underset{\substack{\text{Hydrochloric}\\\text{acid}}}{2\,HCl}$$

The hydrochloric acid is, of course, corrosive, and immediately attacks the iron. It is important to note that the result of this attack is iron chloride, which can break down again as follows:

$FeCl_2$	+	$2\ H_2O$	→	$Fe(OH)_2$	+	$2\ HCl$
Ferrous chloride		Water		Iron hydroxide (turns to rust)		Hydrochloric acid

The free hydrochloric acid is therefore regenerated and the corrosive cycle can begin all over again. This free acid will not combine with any rust or magnesia already present in the boiler water because the boiler conditions will be such that the salts that could be formed are unstable. The free acid therefore continues corroding until it is gradually removed in volatile form with the steam. Even in such a case, however, it may attack the superheaters, the turbines at the low-pressure end, and the economizer.

A similar cycle of reactions occurs with magnesium nitrate as follows:

$Mg(NO_3)_2$	+	$2\ H_2O$	→	$Mg(OH)_2$	+	$2\ HNO_3$
Magnesium nitrate		Water		Magnesium hydroxide (deposit)		Nitric acid

$2\ HNO_3$	+	Fe	→	$Fe(NO_3)_2$	+	H_2
Nitric acid		Iron		Iron nitrate		Hydrogen

$Fe(NO_3)_2$	+	$2\ H_2O$	→	$Fe(OH)_2$	+	$2\ HNO_3$
Iron nitrate		Water		Iron hydroxide (turns to rust)		Nitric acid

Here, again, regeneration of the free acid takes place, and it is to be noted that such regeneration will always take place when there is acid attack of iron in a boiler, because iron salts such as the nitrate, chloride, etc., are unstable under boiler conditions and break down to the familiar rust and the free acid. The hydrogen formed in the action between acid and metal will also tend to cause embrittlement.

Magnesium sulphate itself is quite stable under boiler conditions, but it is not immune from condemnation as above. This is because it may react with sodium chloride, which is usually present in a boiler in appreciable amounts, to give the magnesium chloride according to the reaction:

$MgSO_4$	+	$2\ NaCl$	⇄	$MgCl_2$	+	Na_2SO_4
Magnesium sulphate		Sodium chloride		Magnesium chloride		Sodium sulphate

The magnesium chloride so formed can then decompose as outlined above. This decomposition will facilitate the production of more magnesium chloride since the above reaction between magnesium sulphate and sodium chloride will proceed gradually to completion if the magnesium chloride is removed by decomposition as fast as it is produced. Similar reactions take place with calcium nitrate and chloride, and for convenience the reactions of these acid-forming salts are tabulated below:

$CaCl_2$ Calcium chloride	+	$2\,H_2O$ Water	→	$Ca(OH)_2$ Calcium hydrate	+	$2\,HCl$ Hydrochloride acid
$Ca(NO_3)_2$ Calcium nitrate	+	$2\,H_2O$ Water	→	$Ca(OH)_2$ Calcium hydrate	+	$2\,HNO_3$ Nitric acid

The reactions of the acid with iron, and the subsequent decomposition of the iron compounds to rust and the free acids are as given above, so that the free acid is regenerated in these cases also. It is to be noted that magnesium and calcium hydroxides (or hydrates) are produced in these reactions. These are undesirable in a boiler. The magnesium hydroxide precipitate is very light and flocculent and is very liable to be carried over with the steam if the load is suddenly increased, with consequent ill effect on valves, turbines, etc. The calcium hydroxide may deposit as a hard adherent scale in the boiler.

The Danger of Soft Water

The reactions described above provide an explanation for the mistaken preference still found in some quarters for a few degrees hardness in the boiler feed. This is based on the idea that soft waters are corrosive, and that with a feed of this hardness a protective scale is formed on the boiler heating surface which will protect it from corrosive attack. As explained, many soft waters are undoubtedly corrosive, but distinction must be made between *chemically* softened waters and *naturally* soft waters. The latter should be examined very carefully before adopting as boiler feed. Waters may vary a little from time to time and this variation will be greatest with naturally soft water supplies. Such water may be either slightly acid, slightly alkaline, or perhaps neutral. This is the important point to be remembered when considering the suitability of a soft water supply for the boilers. Soft waters may be dangerous as feed waters, not because they are soft, but because the small amount of dissolved solid which they contain may be acid or may develop acid when the water is used in a steam boiler.

The following analysis of rain water may be considered as an illustration:

Silica	0·02 grains per gallon.
Oxide of iron	0·04 " "
Calcium sulphate ..	0·97 " "
Magnesium sulphate ..	0·40 " "
Ammonium chloride ..	0·08 " "
Sodium chloride ..	0·62 " "
Acidity	0·93 " "
Hardness	1·50 degrees, as parts $CaCO_3$ per 100,000

Although the hardness of this water is low, it contains two undesirable salts. The calcium sulphate would eventually form hard scale in the boiler and even the small amount of magnesium sulphate would probably react in the boiler as described above to give free acid and thus result in corrosion. In addition, of course, there is the appreciable acidity (expressed as sulphuric acid in the analysis) which would cause destruction of the boiler metal.

The same remarks are true of the following analysis of a canal water:

Silica	0·21 grains per gallon.
Calcium carbonate ..	0·87 " "
Calcium chloride ..	0·60 " "
Magnesium sulphate ..	0·51 " "
Magnesium chloride ..	0·29 " "
Sodium chloride ..	3·46 " "
Hardness	3·14 degrees, as parts $CaCO_3$ per 100,000

Here there is both calcium and magnesium chloride which will hydrolyse in the boiler to produce free hydrochloric acid. The magnesium sulphate could also be converted to magnesium chloride and so produce more free acid. This particular analysis shows no free acidity, but where canal waters are used as a source of supply it should be carefully watched since from time to time objectionable effluents are liable to contaminate it. For instance, sometimes water from coal mines is pumped into canals, and such water is usually acid in reaction.

Finally, it should be remembered that natural surface waters are also sometimes very soft, as the following example indicates:

Silica	0·04 grains per gallon.
Oxide of iron	— " "
Calcium carbonate ..	0·46 " "
Calcium sulphate ..	1·28 " "
Calcium nitrate ..	0·07 " "
Magnesium sulphate ..	1·06 " "
Sodium chloride ..	1·97 " "
Hardness	3·41 degrees, as parts $CaCO_3$ per 100,000

The same state of affairs is found here, with the exception that in addition there is a slight amount of calcium nitrate. Such surface waters also frequently contain corrosive organic acids resulting from the decay of vegetation; in addition, the composition of such supplies is very variable.

Sufficient has now been said to show that the danger of naturally soft waters lies in the small amounts of acid-forming substances (and possibly free acid) which they contain. Some treatment should therefore be adopted to militate against the corrosive action of these salts. In this connection it will be interesting to compare the analysis of a lime-soda softened water with the above analyses:

Silica	0·09 grains per gallon.
Calcium carbonate ..	1·37 " "
Magnesium carbonate	1·19 " "
Sodium chloride ..	18·64 " "
Sodium sulphate ..	9·31 " "
Sodium nitrate	3·26 " "
Sodium carbonate ..	2·01 " "
Caustic soda	4·90 " "
Hardness	4·00 degrees, as parts $CaCO_3$ per 100,000

In this case there is alkalinity due to both sodium carbonate and caustic soda, which will counteract any corrosive tendencies due to development of free acid. The absence of these reagents in the above examples of soft water permits any such corrosion to proceed unchecked. It is completely fallacious, therefore, to assume that a lime-soda softener should not be operated to give water of hardness less than 4 or 5 deg. It should be remembered that hardness tests do not indicate whether the salts causing the hardness are scale-forming or acid-forming and the corrosivity of soft waters is due to the presence of small amounts of the latter.

It should not be overlooked that naturally soft waters will also contain dissolved gases, such as oxygen and carbon dioxide, and these are active corroding agents. It is an unfortunate fact that air dissolves most readily in a soft water, so that the naturally soft waters quoted above also contained substantial amounts of dissolved oxygen. This would increase the corrosion caused by the acid-forming salts and should be removed as completely as possible by one of the methods already discussed.

Corrosion by Carbon Dioxide

If the waters contain dissolved carbon dioxide corrosion can then take place according to the following reactions:

Fe	+	CO_2	+	H_2O	=	$FeCO_3$	+	H_2
Iron		Oxygen dioxide		Water		Iron carbonate		Hydrogen

This is then followed by:

$4\,FeCO_3$	+	O_2	+	$10\,H_2O$	=	$4\,Fe(OH)_3$	+	$4\,H_2O$	+	$4\,CO_2$
Iron carbonate		Oxygen		Water		Iron hydroxide (turns to rust)		Water		Carbon dioxide

in which the action of oxygen is seen in another rôle than differential aeration. This is then followed by the breakdown of the hydroxide of iron to rust:

$4\,Fe(OH)_3$	→	$2\,Fe_2O_3$	+	$6\,H_2O$
Iron hydroxide		Rust		Water

Richter has suggested[4] that sludge may form a film which adheres temporarily to heating surfaces and may trap carbon dioxide and hold it in contact with the metal. Corrosion can then proceed more rapidly than when the gas is circulating in boiler water, etc.

It will be observed that the carbon dioxide is regenerated and can therefore initiate another corroding cycle. In this way it can do damage out of all proportion to its amount.

The dissolved carbon dioxide will give the water a slightly acid reaction particularly if it is a soft water such as condensate. The best treatment to correct such acidity is to add lime or lime water to form calcium carbonate, but avoiding excess lime, since its solubility diminishes with increasing temperature and it will therefore scale the heating surfaces. If condensate is being treated, the best procedure is to combine the use of lime and caustic soda. Sufficient lime is used to combine with the carbon dioxide present, and enough caustic soda to give a protective alkalinity. The requisite amounts of these chemicals can be left in the make-up water after treatment and the alkalinity of the boiler water itself can then be controlled by blow-down.

Superheater and Turbine Corrosion

Corrosion due to carbon dioxide and the decomposition of salts under boiler conditions is not confined to the boiler itself. In fact some of the most serious effects are frequently found in the auxiliary plant, the corroding acid or gas more or less " marking its trail " through superheater, turbines, economizer, and condenser. If and when such corrosion occurs it is due to the regeneration which has been pointed out as a characteristic feature of the corroding cycle both with acids and carbon dioxide.

The normal effect of pure steam on the tubes of a superheater is to line the superheater tubes with a film of magnetic oxide of iron, according to the reaction:

$$\underset{\text{Iron}}{3\,Fe} + \underset{\text{Steam}}{4\,H_2O} \rightarrow \underset{\text{Oxide of Iron}}{Fe_3O_4} + \underset{\text{Hydrogen}}{4\,H_2}$$

This film of oxide adheres to the metal and protects it from further attack by pure steam. There may, however, be impurities in the steam which will cause corrosion. These impurities will be of two types, and their presence or absence will be determined by the condition of the boiler water and the nature of the dissolved and suspended solids contained in it. The corrosive impurities may be gaseous or solid ; in the former case they will consist of dissolved gases from the boiler water or volatile acid vapours resulting from the decomposition of salts, and in the latter case will consist of sludge and dissolved salts carried over from the boiler by priming.

Any carbonates present, such as magnesium and calcium carbonates, will decompose at the temperature of the superheater to give carbon dioxide and free lime or magnesia. The corrosive action of carbon dioxide is well known and has already been outlined. The lime or magnesia will be in the form of a fine powder or dust and may be carried over to the turbine. Any unstable salts carried over in priming, such as magnesium chloride, will also be decomposed to corrosive acids. Whether this decomposition occurs in the boiler itself or in the superheater, the result is the same, viz., corrosive attack of the iron. This attack often occurs only at the wet ends, and the reason is that moisture assists in promoting the corrosion and, indeed, is essential in the case of magnesium nitrate. With this salt the acid produced, nitric acid, is dehydrated at a high temperature to give the acid anhydride, N_2O_5, and this does not attack steel, except in the presence of water. Therefore as long as the steam is wet, this anhydride will be present, dissolved in the moisture and will react as acid proper, causing corrosion. When, however, the steam is dried, there will be no moisture present, and the acid will be present as the anhydride only; it will then be innocuous. Such attack will therefore occur at the wet end of the superheater and the low pressure end of the turbines. Scales formed in turbines by corrosive attack of this nature frequently contain considerable quantities of chloride.

Solids may also be carried over into the superheater without any occurrence of priming. This may occur when there is a scum formed on the surface of the water in the boiler; under such conditions a sudden demand for steam will lead to some of this being carried over with the steam. This is particularly likely with magnesium salts. It is often very difficult to remove completely the magnesium salts present in a feed water by lime-soda treatment, and the result may be that an after-precipitation of magnesium hydroxide takes place in the boiler. This precipitate is very light and liable to be carried over with the steam if the load is suddenly increased. In such cases it will usually be found that a turbine deposit contains much magnesium.

Turbine blade deposits may also be formed due to the presence of sodium silicate and/or caustic soda in the solids carried over with the steam. This trouble can be combated by providing certain organic materials or sufficient inorganic reagents, such as sodium sulphate, in the boiler water. The problem is discussed more fully earlier in this book.

Any corrosive attack by carbon dioxide will also occur at the low pressure end of the turbine or the wet end of the superheater. This will be the case because moisture is necessary for the formation of the ferrous carbonate which is the first stage in corrosion by this gas, as shown in the equation above.

A final source of corrosion in both superheater and turbines is any lubricating oil which may have entered the boiler. Such oils are liable to break down under boiler and superheater conditions to give organic acids, and suitable de-oiling arrangements should be provided to prevent such oil from entering the boiler with the condensate or other feed.

Economizer Corrosion

Economizer corrosion is a serious problem, particularly in view of the difficulty in its detection. Either internal or external corrosion may so materially weaken the economizer pipes that a very real risk of serious explosion is introduced. Many economizer explosions, accompanied by loss of life and material damage amounting to many thousands of pounds, have been traced as the direct outcome of corrosion of the pipes.

We are not concerned in this book with external corrosion, but may note briefly that such attack is due to "sweating" of the economizer pipes with moisture condensed from the flue gases. The most effective preventive measure for external corrosion, therefore, is to maintain the feed water temperature at a *minimum of 120° F.*, since this is above the normal dew point of the flue gases.

Internal corrosion is, in some respects, the more dangerous form of attack, since it is often undetected until serious damage or even failure of some of the pipes has occurred. Occasional withdrawal of a pipe or two for breaking up and examination is well worth while to allow some judgment of the extent of any internal corrosion. Such expense may be considered a cheap insurance against an expensive failure, which may be accompanied by loss of life in an economizer explosion.

If the economizer is considered as a kind of feed water purifier, and allowed to remove temporary hardness by decomposition of the bicarbonates, it should not be overlooked that carbon dioxide will be produced. According to conditions, substantial quantities of this corrosive gas may be produced. Serious pitting of economizer pipes has been observed in such cases at points where pockets of gas have been able to collect. Under conditions of thermal decomposition of bicarbonates, the gas is likely to be held in bubbles trapped in sludge or soft scale on the pipe surfaces, as well as collect in bubbles in headers, etc. Dissolved oxygen as well as carbon dioxide may be released by the heat, and cling to top headers in the form of bubbles, which lead to pits. In the case of an untreated feed water, there are also the chlorides and nitrates of calcium and magnesium which may develop acidity in addition to any free acids which may contaminate supplies.

A peculiar form of corrosion experienced in economizers is known as graphitic wasting of the pipes. It owes its name to the fact that after attack the metal has an appearance very like graphite, and can be quite easily cut with a chisel or penknife. Outwardly the pipe seems quite all right, but if it is fractured, or the under layer exposed in some way by scraping, the metal is found to be black and quite soft. It will also soil the fingers just like graphite. In fact, chemical analysis of such black, powdery scrapings shows them to be mainly magnetic or black oxide of iron, with smaller amounts of graphite, silica, and other impurities originally present in the iron.

It will be realized that this form of corrosion is very difficult to detect, the usual measures being to probe and scrape the surface with a penknife, screwdriver, or other small implement. Even better, withdraw a pipe or two and break them up ; the extent of any graphitic wasting is clearly observed by the demarcation of the graphitic layer in the pipe section.

This type of wasting usually occurs as a result of corrosive attack by the dissolved gases contained in the feed water, and is particularly likely to occur when a naturally soft feed water or a large proportion of contaminated condensate is fed to the boilers. The corrosive rôle of both carbon dioxide and dissolved oxygen have already been described elsewhere in this section; the iron carbonate formed in the first case decomposes to the black iron oxide. Trouble has been experienced with economizers operating on water drawn from moorland districts. Such water may at times pass through organic or peaty matter, which is liable to decompose and produce humic acids. These are to some extent soluble in water, and their complex chemical composition makes them unstable. They are liable to decompose under the action of heat—even quite moderate temperatures—giving rise to considerable quantities of carbon dioxide and oxygen. Complete decomposition often produces these gases alone, but partial breakdown may result in formation of simpler organic acids and smaller quantities of gases.

Carbon dioxide can be countered by lime in the ordinary lime-soda treatment. Residual alkalinity in treated water will also take care of any possible acidity or development of acidity in the water. Dissolved oxygen can be removed in a de-aerator, or the water may be passed over iron wool or some chemical such as tannin, sodium bisulphite, etc., will absorb the oxygen.

The Treatment of Alkaline Waters

Corrosion problems such as those described above may develop from the use of alkaline waters, since these contain sodium carbonate with little or no permanent hardness. They usually contain a certain temporary hardness or are soft, but the permanent hardness minerals have been almost completely transformed to the corresponding sodium salts by base exchange action as the waters have percolated through strata of natural sodium aluminium silicates or zeolites. A further difficulty is the danger of caustic embrittlement of the boiler plates if such waters are used as boiler feed untreated. In such treatment a simple reduction of the alkalinity is not all that is desired. Methods must be adopted to prevent *all* the carbon dioxide content and a large part of the sodium carbonate content from reaching the boiler.

Treatment with sulphate of alumina alone is not sufficient. This sets free the carbon dioxide which is combined with the carbonate in the form of bicarbonate. Additional treatment with lime is necessary to remove this gas, otherwise the troubles mentioned will ensue. Great care must also be exercised in adding sulphate of alumina, owing to the acidity of the salt, and too much of it will result in some passing to the boiler unneutralized, where corrosion more severe than that caused by the carbon dioxide gas would result. The ideal treatment if this method is being used is to regulate the chemicals so that (*a*) a slight excess of sodium carbonate is left unneutralized by the sulphate of alumina, and (*b*) sufficient lime is used to take up all the carbon dioxide gas and to convert the above unneutralized sodium carbonate to caustic soda. This slight amount of caustic soda will not give rise to embrittlement, and danger from carbon dioxide and disintegration of sodium carbonate will be eliminated.

Analogous to the sulphate of alumina treatment is the more recent method of neutralizing the alkalinity with free sulphuric acid. The free carbon dioxide resulting from the interaction between the bicarbonate and the acid can be largely removed by passing steam into the water. In such a case, some slight amount of unneutralized bicarbonate or carbonate might be left in the water after adding acid, then removing the carbon dioxide with steam, and finally treating with lime to take care of any traces of the gas which may be left and also to convert the carbonate left to caustic soda. Treatment on these lines is adopted for zeolite softened water containing too much bicarbonates at the Commonwealth Edison Co.'s Chicago power station.

Alkaline water may also be successfully treated by the reverse of the lime-soda treatment. In ordinary lime-soda softening the permanent hardness is removed with sodium carbonate. Therefore the carbonate in alkaline waters can be treated with any permanent hardness salt. A suitable combination is calcium chloride or sulphate with lime. The reactions may be carried out in any modern type of lime-soda softener, according to the following reactions:

(i)	$Na_2CO_3.H_2O.CO_2$	+	$Ca(OH)_2$	=	Na_2CO_3	+	$CaCO_3$	+ $2\,H_2O$
	Sodium bicarbonate		Hydrated lime		Sodium carbonate		Calcium carbonate (deposit)	Water

(ii)	Na_2CO_3	+	$CaSO_4$	=	Na_2SO_4	+	$CaCO_3$
	Sodium carbonate		Calcium sulphate		Sodium sulphate (soluble)		Calcium carbonate (deposit)

and, to convert the slight amount of unneutralized carbonate to caustic soda:

(iii)	Na_2CO_3	+	$Ca(OH)_2$	=	$2\,NaOH$	+	$CaCO_3$
	Sodium carbonate		Hydrated lime		Caustic soda		Calcium carbonate (deposit)

There are therefore three final products—calcium carbonate, sodium sulphate, and a small amount of caustic soda. The carbonate settles out and can be filtered off, the sodium sulphate is not dangerous unless the concentration becomes high, and the slight amount of caustic soda will give the boiler water the necessary slight amount of alkalinity to combat corrosion, etc. The sulphate concentration and degree of alkalinity may be adjusted when required by blow-down. If the treated water is being used with condensate a slight excess of lime in the water is also advantageous to take care of any carbon dioxide in the condensate. Any temporary hardness in the water must, of course, be removed with lime.

A recent development is the hydrogen zeolite method of treatment, which removes carbonate and bicarbonate of sodium completely and thus enables an effluent water to be obtained with any desired alkalinity. The method is discussed fully in preceding articles.

Another expedient for dealing with alkaline waters is the rarely practised procedure of blending. The blending of two or more different waters may lead to substantial saving in the chemicals required for treatment of any of the waters. This is effected owing to the different character of the saline concentration of the waters and the procedure to be adopted is described in detail earlier in the book.

It may be of interest before leaving this subject to point out that alkaline waters may be used in connection with the de-oiling of water. Condensate may become contaminated with oil, and the disastrous effects of this in the boilers need no emphasis. A good way of removing the oil is to treat the water with sodium carbonate and sulphate of alumina. It is then removed by becoming entangled in the gelatinous precipitate of aluminium hydroxide which is formed. The sodium carbonate in an alkaline water may therefore be used in this process. The water will, of course, require additional treatment for the removal of carbon dioxide, etc., as mentioned.

Caustic Embrittlement

The cracking of boiler plates known as embrittlement is one of the most serious problems of the modern boiler-house engineer. It occurs in the riveted areas of the boiler, the cracks passing from one rivet hole to another, and in fact are found at those places likely to be subject to the highest localized stresses, such as in drum seams, near rivets, and at tube joints. The cracks often run past one another and are in fact irregular in direction, but never extend into the body of the plate beyond the lap of the seam ; they have been found in different types of boilers, low as well as high pressure, and in plates of good as well as poor quality. The fracture resembles that of a brittle metal more than that of a normal ductile boiler plate. Hence the term embrittlement. Microscopic examination shows that the boiler steel has good crystal structure, the carbon being well distributed in even-grained pearlitic form. By means of extreme cold work the ductility of a boiler plate may be lowered and the metal become brittle. However, in such a case the cracks are transcrystalline. Micrographic study of embrittlement cracks does not show any sign of this ; here the cracks are intercrystalline—they follow the grain boundaries.

Numerous and conflicting theories have been put forward to explain this type of failure. It has been contended that it is brought about by poor metal, excessive cold work, too high a riveting pressure, improper design of boiler, etc. In recent years the generally accepted explanation has been that it is due to the presence of a caustic alkaline content in the water. The fact that concentrated caustic soda at boiler temperatures attacks iron with the evolution of hydrogen, coupled with the knowledge that hydrogen generated in contact with iron causes the metal to lose its ductility, led the industry to say the hydrogen had embrittled the boiler plate. Therefore the term " caustic embrittlement," meaning a loss of ductility caused apparently by hydrogen resulting from caustic action, was applied to this behaviour. However, the suggestion was put forward by some authorities that the effect was largely physical, and that the explanation was to be found in the continuously repeated stresses set up in the plant by water surging and similar causes.

As a result of a vast volume of research work (a comprehensive review has just been published[5] with voluminous references), it seems likely that the various theories are partly correct, though the matter is still not fully understood. To-day the general consensus of opinion is that three conditions must simultaneously exist before embrittlement can occur:

(1) Suitable chemical conditions in the boiler water.

This means the presence of caustic soda, the absence of inhibiting agents, and (at higher pressures particularly) the presence of other chemicals, such as silica, which assist in localizing and accelerating attack at the grain boundaries.

(2) Localized production of high concentrations of the chemicals.

Laboratory and practical tests have demonstrated how this occurs in small crevices in boiler seams, under rivet heads, etc.

(3) High metal stresses in the region of the concentrated solution.

It has been estimated that the actual stress in a boiler plate at a rivet hole is nearly three times the average stress in the plate, or approximately equal to the yield point strength of the plate. In addition, there are stresses caused by caulking, by rivet holes not being absolutely in line, etc., so that local stresses about equal to the yield point strength probably exist. It is reasonable, too, to conclude that any such stresses in the metal are increased by operating conditions and the continuous surging of the boiler water.

These factors have become recognized as a result of numerous investigations on plant as well as laboratory scale, researches in recent years having been greatly assisted by the development of embrittlement detectors. The Schroeder and Berk detector[6] is sketched at Fig. 21. It is simply a device which enables it to be determined whether or not a given water is of embrittling character. The apparatus can be attached directly to a boiler so that a portion of the boiler water continuously pours into or through it and thus avoids the necessity of sampling with its attendant difficulties and errors. The specimen steel is held in place by the clamping block and nuts A. An adjusting screw enables the investigator to establish the desired leak and at the same time to stress the specimen. The continuous but minute leakage to the atmosphere so arranged causes the water, by evaporation, to reach a high degree of concentration. The concentrated solution is then in contact with the stressed area of the specimen. The embrittling characteristics of the water make up the third factor under consideration in the test. The feed lines may be fitted with valves to allow the replacement of specimens while the boiler is under full pressure, or of course the apparatus may be attached to a small laboratory boiler or autoclave for laboratory tests. With waters of an embrittling nature, typical intercrystalline cracks occur in the specimen usually within thirty days. The Straub and Bradbury detector[7] is suitable for laboratory tests and subjects prepared steel specimens to high pressures and temperatures in concentrated caustic solutions in a small steel autoclave while under mechanical tension.

Although the conditions under which cracking takes place are known,

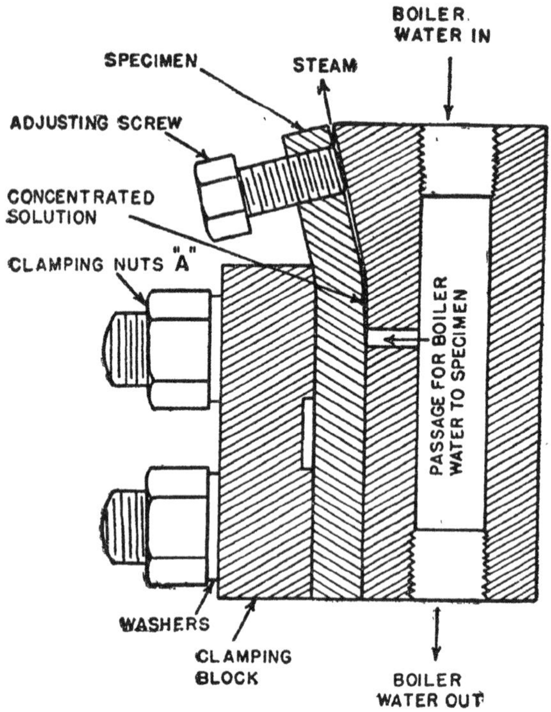

FIG. 21. Embrittlement detector (Schroeder and Berk).

the exact mechanism of the attack is still not fully understood. It now seems probable that it is a process of selective corrosion on the following lines. The first step is the leakage of caustic soda solution into the underwater seams and their evaporation there to high concentrations. It is easy to visualize concentration in various small crevices, capillaries, etc., which leak sufficiently to allow steam to flash out to the atmosphere and are yet small enough so that there will not be such a flow of water as to prevent the build-up of strong solutions in contact with the metal. If this action continues over long periods, very high concentrations will build up, investigations having shown in some cases[8] concentrations

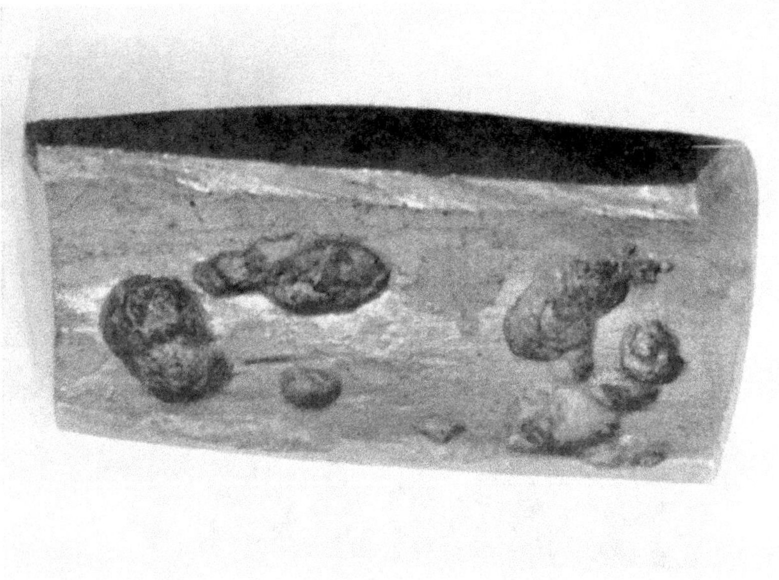

PLATE 15. CORRODED ECONOMISER TUBE
(*Boiler House Review*)

PLATE 16. WEIR "OPTIMUM" DE-AERATOR INSTALLED IN WORKS POWER PLANT
(*By courtesy of Huntley & Palmers, Ltd., Reading*)

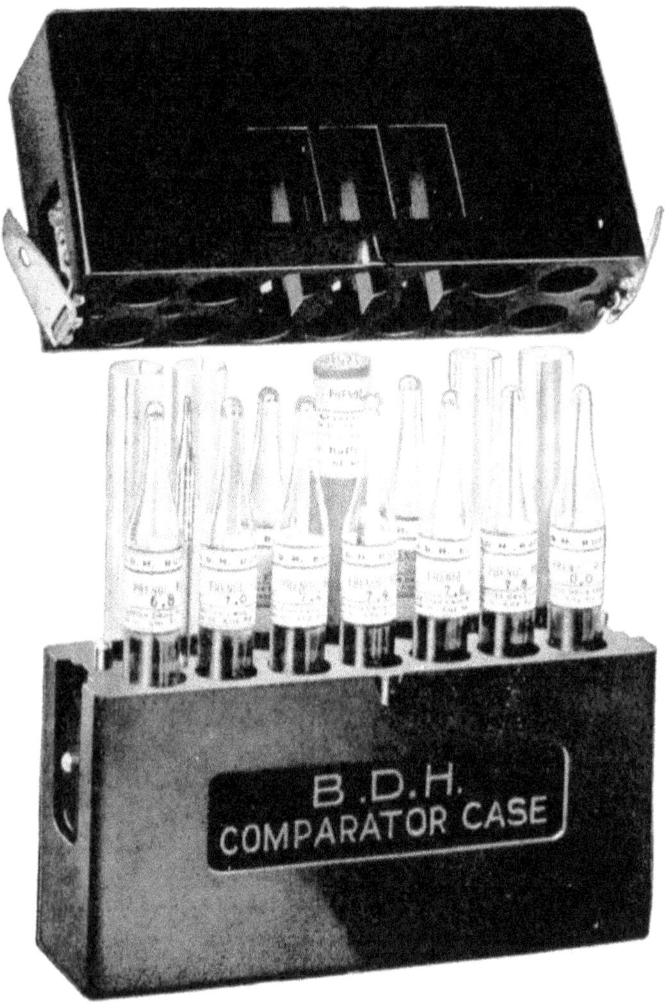

PLATE 17. FORM OF HYDROGEN ION COMPARATOR

approximating 1,200 times that carried in the average blow-down water. High caustic soda concentrations are thus produced in regions of high metal stress. The chemical action from this stage is uncertain and possibly a complex one involving several factors, but it is believed that there is a basic reaction:

$3\,Fe$	+	$4\,H_2O$	=	Fe_3O_4	+	$4\,H_2$
Iron		Water		Iron oxide		Hydrogen gas

One theory is that this process is accelerated by high concentrations of caustic soda, since the latter increases the electromotive force of the solution with respect to the steel, and the reaction is progressively accelerated if the concentration steadily builds up as described. Furthermore, an increased E.M.F. is to be expected at the grain boundaries owing to the boundary atoms being in a state of strain. Another idea is that at high concentrations the caustic soda dissolves the film of oxide formed on the metal, and thus permits the reaction to continue. More rapid action at the grain boundaries is explained by the fact that they are more active chemically owing to the energy stored up there by stress—or, in other words, atoms at the grain boundaries are in a state of strain and consequently more readily removed than atoms within the crystals of the steel. The structural stresses at seams, etc., still further increase the strain and consequently the chemical activity of atoms at the grain boundaries. It is thought that the rôle of silica is to aid under certain conditions in establishing a protective film on the crystal faces, thus leaving the grain boundaries more open to attack. Finally, the oxide Fe_3O_4 produced occupies a greater volume than the original metallic iron, so that great expansive forces are generated. The metal therefore cracks, particularly since it is already stressed to near its yield point in these regions. Another theory[9] is that the atomic hydrogen resulting from the reaction of iron with the water forms gas pockets which exert tremendous pressures to separate the grain boundaries by cracking.

Preventive measures developed as a result of an extensive investigation of U.S.A. and European cases[10] consisted in maintaining in the boiler water the following sodium sulphate/total alkalinity ratios:

1	for boilers operating up to 150 lb. pressure	
2	„ „	from 150–250 lb. pressure
3	„ „	over 250 lb. pressure

These results were in accord with a considerable number of observations on the subject in Europe. For instance, Splittgerber,[11] in Germany, suggested minimum sulphate/alkalinity ratios which varied from 3 at 250–350 lb. pressure to 8 at 750–850 lb. pressure. Another German worker, Berl,[12] came to similar conclusions to those reached in America, and it is interesting to note that his experiments also confirmed the possibility of local concentration of the boiler water in seams, hair cracks, and capillaries as described. This preventive action has been attributed to the ability of sodium sulphate to lower the E.M.F. of boiler water with respect to steel (and therefore reduce or stop the reaction set out above), and to a physical sealing of crevices from access to the caustic boiler water by the prior deposition of sodium sulphate, since this mineral is much less soluble than caustic soda at boiler temperatures.

For some time these sulphate/alkalinity ratios appeared to be the answer, since early research work was supported by favourable practical experience, but data gradually accumulated to show that sodium sulphate is not a *certain* inhibitor of embrittlement, particularly at pressures above 250 lb. and in railway practice.[13] It has been suggested[8] that such conflicting experiences are due to different physical conditions which in some cases may militate against the adhesion of a sealing deposition. It is probable, too, that varying chemical conditions in the boiler waters contribute to these different operating results; various recent research programmes have revealed inhibiting effects exercised by unexpected organic and inorganic compounds, and the presence of any of these may therefore have been responsible for the apparently contradictory observations.

Further research work[14] indicated that silica takes part in the embrittlement attack, particularly at high pressures, and improved methods of prevention were put forward as follows[15]:

Up to 250 lb. pressure ...	Maintain sodium chloride/total alkalinity greater than 0·6 *And* sodium sulphate/total alkalinity greater than 1·0
Pressure 500–1,400 lb. ...	Maintain in boiler water soluble salts containing iron and aluminium oxides at concentration greater than 0·6 times concentration of silica present

Either method was recommended for the intermediate pressure range, and again early operating results for a number of plants[16] showed some support for the new controls.

The most recent laboratory and plant trials,[13] however, have shown that embrittlement may still not always be prevented by these new methods, and new inhibitors have been examined as follows:

Sodium Nitrate.—Very good laboratory results have shown it to be effective up to 400 lb./sq. in. if ratio sodium nitrate/total alkalinity is not less than 0·4. Locomotive boiler experience has been good but limited. A promising inhibitor, but plant experience is required. It is an inexpensive chemical, easily handled and controlled, and it is known that it is thermally stable and will not interfere with other boiler water chemicals.

Tannins.—Certain tannins, especially the quebracho type, have been found effective with a tannin/total alkalinity ratio of 0·4. Usual limit is 500 lb. pressure, but satisfactory results have been reported in one case at 650 lb.

Lignins.—Lignin-sulphonates contained in waste sulphite liquors have been used successfully on many locomotive boilers, but there is risk of interference with other substances in boiler waters, causing much foaming, so that they are not recommended at present for stationary boilers. A combination of these materials with sodium nitrate is being investigated.

Phosphates.—Claims that sodium phosphate is an effective inhibitor[17] have been disproved by German[18] and American[13] investigations. These attempts to use phosphates should not be confused with the alkalinity control method below.

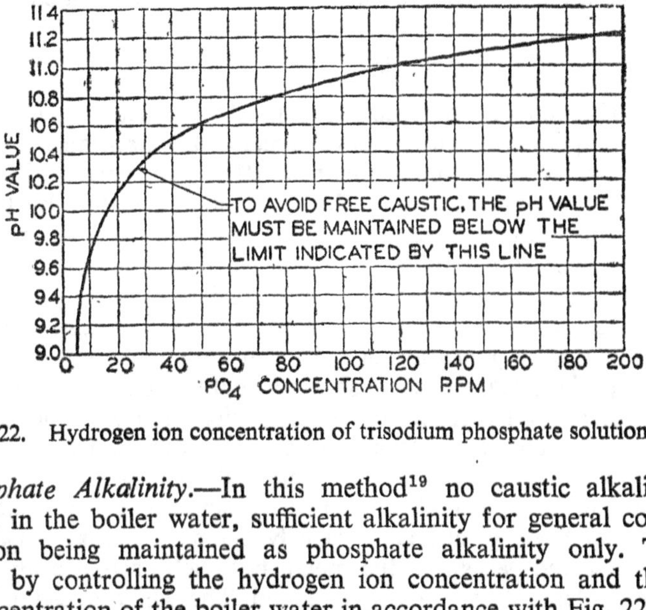

FIG. 22. Hydrogen ion concentration of trisodium phosphate solutions.

Phosphate Alkalinity.—In this method[19] no caustic alkalinity is allowed in the boiler water, sufficient alkalinity for general corrosion inhibition being maintained as phosphate alkalinity only. This is effected by controlling the hydrogen ion concentration and the PO_4 ion concentration of the boiler water in accordance with Fig. 22, which shows the relation between pH and PO_4 concentration in pure trisodium phosphate solutions. A point which falls below the curve shows the presence of acid phosphate salts; a point in the region above the curve is evidence that caustic is present. Therefore as long as the pH and phosphate concentration conform to this curve evaporation will deposit trisodium phosphate and no caustic soda The optimum condition, with respect to embrittlement, corrosion protection, and boiler water salines is the maintenance of the pH value slightly below that on the curve corresponding to the desired phosphate concentration. If oxygen scavenging is needed, sodium bisulphite can be used effectively to reduce the pH without disturbing the PO_4 concentration. The use of phosphoric acid is particularly recommended where the feed water is alkaline or contains decomposable sodium carbonate, but any other mineral acid may be used if desired to reduce alkalinity without increasing the phosphate concentration. Experience with this new technique is limited (laboratory tests were done up to 800 lb. pressure), but has been reported to have been found successful (as measured by the detector) in four plants.[20] It is claimed to be the most practicable way to prevent embrittlement in high-pressure plants using little make-up water.

In the present state of our knowledge, it would be unwise to rely too much on any ratios so far specified for prevention of embrittlement. If confronted with this problem, the best plan probably is for the operator to use an embrittlement detector to determine the embrittling properties of the water actually in use in his boilers and to check the efficiency under his own conditions of any preventive treatment contemplated.

Hydrogen Ion Concentration

The introduction of hydrogen ion control in boiler feed water problems is a comparatively recent development, its importance being first pointed out by Wolman and Hannan in 1921. Primarily it was used in the control of clarification of turbid and coloured waters by the addition of iron or alum salts, or both, but in the case of boiler feed water it is best known from the fact that the establishment of an optimum hydrogen ion concentration diminishes corrosion.

The subject is a closed book to many, chiefly because there is an impression that the application of hydrogen ion control requires the services of a specially trained chemist. As a matter of fact, it is almost as simple to determine hydrogen ion concentration as it is to measure temperature, and these determinations can be made practically as easily and accurately by a plant foreman as by a trained chemist.

It has already been stated that corrosive action between boiler water and the metal heating surface is largely chemical action between the salts and dissolved gases and the metal, but such corrosive attack may be largely accelerated by electrolytic action. The whole question of such interaction is closely bound up with what is known as the "reaction" of the water; in other words, whether the water is acid, neutral or alkaline. Corrosion is very rapid and severe in an acid solution, but may be stopped completely when the water has a certain degree of alkalinity. At one time the only way to find out the reaction of a sample of water was to place a piece of litmus paper in it, and if the colour became red the water was acid, whereas a blue colouration indicated alkalinity. This, however, did not indicate the actual degree of alkalinity, so that there was no simple way of saying if the water was sufficiently alkaline for any given purpose. Nowadays the hydrogen ion concentration may be rapidly and accurately determined and recorded, and as this gives an exact measure of the acidity or alkalinity, working conditions can be standardized to the optimum point.

The idea that in the process of electrolysis a compound is partially dissociated into *ions* is familiar to most engineers, particularly in connection with accumulators. During electrolysis these ions or minute particles of matter travel from one pole of the accumulator to the other, and the Greek origin of the word ion means wanderer or traveller. According to the theory of electrolytic dissociation, all liquids which contain water also contain free hydrogen ions (H^+) and hydroxyl ions (OH^-). When the number of hydrogen ions exactly equals the number of hydroxyl ions the solution is said to be neutral. If the number of hydrogen ions exceeds that of the hydroxyl ions the solution is said to be acid, while if there is an excess of hydroxyl ions it is said to be alkaline.

Experiments have shown that pure water dissociates slightly into hydrogen and hydroxyl ions, there being 1 gram of hydrogen ion in each ten million litres of water. Since pure distilled water is neutral it follows that there must be the above amount of hydroxyl ion also present. In other words, the concentration of each is one ten-millionth of a gram per litre, or 1×10^{-7} gram per litre. This means that the product of the concentrations of these two ions is $C_H \times C_{OH} = 10^{-7} \times 10^{-7} = 10^{-14}$. According to a certain chemical law known as the law of mass action, this product of the concentrations of the ions is a constant at any given temperature, and is known as the dissociation constant of water. (Incidentally, this is why it is usual to speak only of the hydrogen ion concentration since if one is known the other is readily ascertained.)

This method of expressing hydrogen ion concentration is very inconvenient, and it is now customary to quote a "pH value." This is the logarithm to the base 10 of the reciprocal of the hydrogen ion concentration in grams per litre. Therefore $pH = \log \frac{1}{[H^+]}$ where $[H^+]$ is the concentration of hydrogen ions in grams per litre; this is simply the index of the hydrogen ion concentration with the negative sign changed to positive. Therefore the pure water with a hydrogen ion concentration of 10^{-7} has a pH value of 7·0. It has already been stated that in pure water the hydrogen and hydroxyl ions exactly balance one another. This value of 7·0 is thus the neutral point on the pH scale, the scale actually covering all possible degrees of acidity and alkalinity. In fact, the pH scale may be interpreted as follows:

When pH = 7·0 the feed water is neutral.
When pH is *less* than 7·0 the feed water is *acid*.
When pH is *greater* than 7·0 the feed water is *alkaline*.

The actual range extends from just below zero to 14·5, i.e., equidistant each side of the neutral point.

When acids are added to water they dissociate to a certain extent to give hydrogen ions and acidic radical ions. A corresponding decrease in hydroxyl ion concentration must therefore take place. Similarly, when an alkali is added to water it increases the amount of hydroxyl ions and causes the hydrogen ion concentration to diminish. The extent of such an increase depends on the amount and strength of the acid or alkali which is added. Acids or alkalis are classified according to their "strength" or degree of dissociation into ions. A strong acid is largely dissociated to give hydrogen ions; for example, decinormal hydrochloric acid is dissociated over 90 per cent, while the same strength acetic acid (a weak acid) is only 1·3 per cent dissociated. This means that although a given amount of each of these solutions will require the same amount of alkali for neutralization, yet the *active* acidity of the hydrochloric acid will be some 70 times as much as that of acetic acid. The pH value, therefore, does not measure the amount of acid present so much as its intensity or activity.

Another factor is buffer action. This operates similarly to a weak acid above—it reduces the hydrogen ion concentration. Certain salts, though they may be neutral in themselves, can dissociate into ions, and these ions can exert a neutralizing effect on any acid or alkali which is added. An example is the case of a solution of a mixture of acetic acid and sodium acetate. In this solution the dissociation of the acetic acid (a weak acid) is very small, thus:

$$\underset{\text{Acetic acid}}{CH_3COOH} \rightleftarrows \underset{\text{Acetate ion}}{CH_3COO^-} + \underset{\text{Hydrogen ion}}{H^+}$$

The dissociation of sodium acetate, on the other hand, is large:

$$\underset{\text{Sodium acetate}}{CH_3COONa} \rightleftarrows \underset{\text{Acetate ion}}{CH_3COO^-} + \underset{\text{Sodium ion}}{Na^+}$$

If a small quantity of hydrochloric acid is added to this solution it is immediately dissociated into its ions:

$$\underset{\text{Hydrochloric acid}}{HCl} \rightleftarrows \underset{\text{Hydrogen ion}}{H^+} + \underset{\text{Chlorine ion}}{Cl^-}$$

There is then in solution hydrogen, chlorine, sodium and acetate ions. Since acetic acid ionizes only to a slight extent, the hydrogen and acetic ions cannot exist together in solution to a very large extent. The hydrogen ions from the hydrochloric acid will therefore immediately combine with the acetic ions (CH_3COO^-) to form undissociated molecules of acetic acid, and the pH of the solution will then remain unchanged. The hydrochloric acid added would, of course, be shown by titration with alkali, but it would not be in an active condition in the solution. Now suppose, instead of acid, that caustic soda is added to the solution. This should increase the hydroxyl ion concentration and therefore diminish the hydrogen ion concentration. Actually the caustic will react with the hydrogen ions to form sodium ions and undissociated water molecules. More acetic acid will then dissociate into H^+ ions and CH_3COO^- ions since the product of the concentrations of these two ions has been taken below the equilibrium point by the removal of hydrogen ions by the caustic addition. The hydrogen ion concentration will again be unchanged. This is buffer action by sodium acetate, and considerable amounts of acids and alkalis can be added to solutions containing such salts without changing the pH value. Essentially, therefore, such salts have a neutralizing effect on acids and alkalis and prevent hydrogen ion concentration from being changed. Another consequence of such action is that buffered solutions can be diluted with distilled water, even though the water may show a very acid reaction, without affecting the pH value. In fact, some solutions can be diluted to hundreds of times their original volume. This is often important in making determinations on very highly coloured and turbid solutions.

Buffer action is also of importance in providing standards for colorimetric determinations of pH values. Owing to this action, solutions can be made up which have definite hydrogen ion concentrations, and, by buffering them, it can be ensured that the hydrogen ion concentration will not change with any slight variations in acidity or alkalinity which may occur. Then, by addition of a certain indicator to this solution, a colour standard is created. If an unknown solution gives the same solution with this particular indicator, then its pH value must be the same as that of the standard. In general, the salt of any weak base or weak acid is a buffer salt, so that there are very few solutions which are free from buffer action. Thus, carbonates in raw water have buffer action. This is fortunate, since it will be appreciated that if considerable amounts of free acid or alkali can be added to such buffered solutions without affecting their pH value, no special precautions are necessary in making pH measurements on buffered solutions.

Determination of pH Values

There are two methods of measuring the pH value of a solution, the electrometric and the colorimetric. Colorimetric measurements of pH are based on the principle that various indicators change in colour when acted upon by different solutions of various acidities or alkalinities. A very large number of indicators used in determining acidity owe their utility to the fact that they undergo a change in colour within a certain range of hydrogen ion concentration, and can therefore be used to determine the hydrogen ion concentration in this range. Thus, phenolphthalein is colourless in solutions with a pH greater than 8, pink at pH 9, and deep red at pH 10. Litmus paper is another indicator—probably the simplest and best known. Its colour is red in acid solution and blue in alkaline. If the paper turns slightly red, the solution is said to be " slightly " acid, while if it turns deep red the solution is supposed to be " strongly " acid. Similarly, there is " slightly " and " strongly " alkaline. These are very indefinite terms and their meaning varies enormously, depending largely on the personal opinion of the worker. Thus, what is slightly acid to one man may be strongly acid to another. By making pH measurements definite pH values are substituted for the terms "strongly" and "slightly" alkaline or acid, and these values can be recorded and duplicated at any time by any operator.

Theoretically, litmus and other test papers can be used for determining the pH of a solution. In practical application, however, the error is too great, chiefly because the range is too wide (pH 4·6 to 8·4) for the colour changes to be distinct. It is practically impossible to tell the difference in colours produced by solutions having pH values which differ by one unit. Some indicators do not cover acid and alkali ranges, but change their colour with change of acidity alone.

In order to make accurate pH determinations, it is therefore necessary to employ indicators with short ranges, so that the colour changes are more distinct. This means that a number of indicators are required to cover the range pH 1·2 to pH 12·6. A complete set of indicators, developed for just this purpose, is given in Table 10.

TABLE 10

Indicators for Hydrogen Ion Concentration Values

Indicator	pH Range	Colour change
Metacresol purple	1·2— 2·8	Red—yellow.
Bromphenol blue	3·0— 4·6	Yellow—blue.
Bromcresol green	4·0— 5·6	Yellow—blue.
Chlorphenol red	5·2— 6·8	Yellow—red.
Bromthymol blue	6·0— 7·6	Yellow—blue.
Phenol red	6·8— 8·4	Yellow—red.
Cresol red	7·2— 8·8	Yellow—red.
Thymol blue	8·0— 9·6	Yellow—blue.
Nitro yellow	10·0—11·6	Yellow—orange.
Sulpho-orange	11·0—12·6	Yellow—orange.

The colour changes of these indicators are similar to that of litmus. For example, litmus is deep red in acid solution (pH 4·6) and deep blue in an alkaline solution (pH 8·4). Similarly, chlorphenol red has a yellow colour at pH 5·2, and a deep red colour at pH 6·8 ; bromthymol blue is yellow at pH 6·0, and deep blue at pH 7·6, etc.

If the pH is not known even approximately, it is necessary first to make a rough determination by testing with a few different indicators. To make the test, fill three or four graduated test tubes to the mark with the solution to be tested. Each test tube should contain 10 ml. of solution. To the first one add 0·5 ml. of bromthymol blue indicator solution. This indicator is tried first since its pH range (6·0 to 7·6) covers the neutral point pH 7·0. We are therefore able to say at once whether the solution being tested is neutral, acid, or alkaline. The colour change of bromthymol blue is from yellow to blue, so that if a colour intermediate between yellow and deep blue is obtained with the solution, then its pH is between 6·0 and 7·6, and it is either neutral or slightly acid or alkaline. In this case no further rough tests are necessary.

If, however, a yellow colour is obtained on adding the bromthymol blue indicator solution, the pH of the solution is at least 6·0 and possibly lower, since pH 6·0 is the acid end of the range of this indicator. Even if the solution had a pH of 5·0 or 4·0 it would still turn bromthymol blue indicator yellow. Therefore, if a yellow colour is obtained, the next indicator must be tried in the same way, i.e., 0·5 ml. of bromcresol green is added to the second test tube containing the solution under test. The colour change will then tell us whether the pH of the solution lies in the bromcresol range (pH 4·0 to 5·6).

By systematic procedure in this way one can quickly determine the range which covers the pH value of the solution. The next thing is to find the exact pH value, but before describing how this is done, the "universal" indicator deserves mention. This ingenious device is a mixture of several indicators so designed that addition of a few drops to a solution gives a colour characteristic of the pH, the succession of colours according with that which is found in the spectrum:

TABLE 11

Universal Indicator Colour Changes

pH	Colour	pH	Colour
3	.. Pale-red.	7–7·5	.. Greenish-yellow.
4	.. Red.	8	.. Green.
5	.. Orange-red.	9	.. Greenish-blue.
6	.. Orange-yellow.	10	.. Violet.
6·5	.. Yellow.	11	.. Reddish-violet.

This may be used for obtaining an approximate idea of the pH. Thus, if the universal indicator has given a reddish-orange tint, select an indicator covering the range pH 4–6 and use buffer standards of the indicator over that range to determine the pH accurately. Such a universal indicator is marketed by The British Drug Houses, Ltd.

Having found by rough tests the pH range which covers the pH of the solution, a set of standards should be obtained of some indicator which covers that range, and also some form of comparator. Suppose the pH lies between 6·0 and 7·6; the indicator to be used would be bromthymol blue. A set of bromthymol blue colour standards consists of nine standard tubes and one tube of distilled water. The nine tubes contain 10 ml. of solutions of definite pH value—that is, 6·0, 6·2, 6·4, 6·6, 6·8, 7·0, 7·2, 7·4, and 7·6—to each of which 0·5 ml. of bromthymol blue indicator solution has been added. In short, these are buffer standards of the indicator which have been made up in exactly the same manner in which the test was made. Of course, owing to their buffering properties, these standards are of definite and unchanging value, otherwise they would be useless for comparison. The tube marked pH 6·0 will thus have a yellow colour, and the one marked pH 7·6 a deep blue colour, while the intermediate tubes will have colours between yellow and blue, a gradual transition from yellow to blue being observed.

It will be obvious that to determine the pH accurately the test sample must be matched with the standards, and for this purpose a comparator arrangement, illustrated in Fig. 23 and Plates 17 and 19, is used. Such an arrangement is necessary in order to eliminate the effect of colour and turbidity of the sample when making comparisons.

Three test tubes are filled to the mark (10 ml.) with the solution to be tested, and are placed in three blackened holes bored into a block of wood. (Holes A, B, and C.) In A and C samples of the solution no indicator is added, but to tube B is added 0·5 ml. of bromthymol blue indicator solution. This is done by means of a small graduated pipette supplied with the comparator block. The tube is shaken to ensure thorough mixing. In D and F we insert standard buffer solutions of the indicator which differ by only 0·2 pH—for example, in this case, say, the two bromthymol blue colour standards 6·8 and 7·0. Finally, the hole E is filled with a tube of distilled water. The comparator block

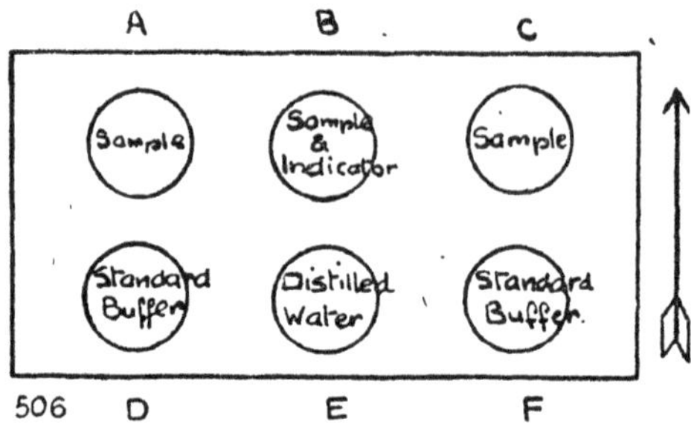

FIG. 23. Comparator arrangement for pH determination.

represented by Fig. 23 contains three slots, extended through the block from D to A, E to B, and F to C, so that the tubes can be compared in pairs.

The reason for this arrangement of tubes is to eliminate any effect of colour or turbidity in the sample which is being tested. Take each pair of tubes in turn. The standard buffers may be considered as made up of water and indicator solution. They do, of course, contain solutions of salts to give them definite pH value, but as these solutions of salts are clear and colourless, they may be considered as made up of only these two materials. Therefore, each pair of tubes, A and D, B and E, C and F, contains the sample of the solution being tested, the indicator solution, and water. In short, we have the same combination in all pairs, so that, when the proper colour standards are in place an exact match is obtained, since the observer is looking through exactly the same materials in each pair of tubes. The match is made as follows: Look through the three pairs of tubes, holding them to the light, and changing the colour standards if necessary, until the central pair of tubes exactly matches one of the other pairs, or until the colour is intermediate between the colours of the pairs on either side. In the former case the pH can be read off directly from the standard with which a match is obtained; in the latter, the pH is the average of the two standards. A few precautions should be mentioned. The standard buffers in D and F should never differ by more than 0·2 pH. When matching colours, the worker should always look through the slots in the direction of the arrow. Again, the pH ranges of practically all the indicators overlap, so that one can make determinations on a given sample with two different indicators. Such procedure is a valuable check on results obtained. It is also advisable where a test sample matches the colour standard on either end of a given set. As already mentioned, a test sample may match, say, the bromthymol blue standard pH 7·6, and yet have a much higher pH value, pH 7·6 being the end of the range of that indicator. Therefore, the test should be repeated using another indicator which overlaps.

The conditions would be similar if the solution being tested was turbid instead of coloured. Special comparators for such cases are now on the market. In any case, this difficulty can often be overcome by dilution or by filtering through a good grade of wood charcoal. The charcoal is first washed with alkali, then acid, followed by distilled water. If this is done the pH of the solution will not be changed to any great extent, especially if it is highly buffered.

Hydrogen Ions and Corrosion

The corrosion of iron in boilers may be represented as an electrochemical process proceeding on the following lines. All metals have a tendency to go into solution when in contact with water or aqueous solutions, and the process is represented as an electrical interchange thus:

Fe	+	$2H$	→	Fe^{++}	+	H_2
Iron (metal)		Hydrogen (ions)		Iron (ionic form)		Hydrogen (gas)

The iron metal goes into solution as a result of the interchange of electrical charges from the hydrogen ions in solution. The iron then exists in solution as electrically charged particles or ions, and the hydrogen from which the charges are taken is released as a gas. The extent to which the above action takes place is very small, unless various other conditions are established which result in the removal of the iron from solution. Under such conditions, the solution of metallic iron, or in other words its corrosion, steadily continues as long as the ionic particles in turn are thrown out of solution in some form such as rust.

As already explained in Part 2 of this book, water is slightly dissociated into ionic particles:

H_2O	⇆	H^+	+	OH^-
Water molecule		Hydrogen ion		Hydroxyl ion

The hydrogen ions which participate in the above solvent or corrosive action are therefore present in all aqueous solutions to an extent which depends upon local conditions such as the temperature to which the water is exposed. Furthermore, after conversion to hydrogen gas the hydrogen ions leave behind an equivalent number of hydroxyl ions. These balance up with the ferrous ions which pass into solution, so that the solution or water remains electrically neutral, as follows:

Fe^{++}	+	$2\,OH^-$	$\rightleftharpoons$	$Fe(OH)_2$
Iron ions		Hydroxyl ions		Iron (ferrous) hydroxide

In short, the ferrous ions and the hydroxyl ions tend to unite to form ferrous hydroxide, and again a state of balance is set up so that at a

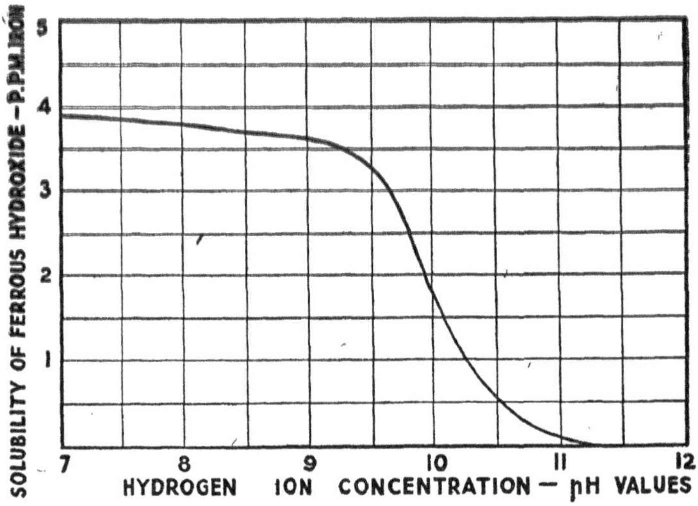

FIG. 24. Effect of alkalinity on solubility of ferrous hydroxide.

given temperature the product of the concentration of the ferrous and hydroxyl ions is a constant.

The solution or corrosion of iron in this way will cease after proceeding to only a very slight extent normally owing to (*a*) the accumulation of a minute film of hydrogen gas on the metal surface, thus preventing contact with the water and hydrogen ions; and (*b*) owing to the iron concentration in the water reaching the saturation point, plus a tendency for a protective oxide film then to be deposited on the metal surface. The presence of oxygen dissolved in the water, however, will upset these protective influences. The hydrogen will combine with the oxygen to form water, thus maintaining metal/water contact. The oxygen will also combine with the ferrous hydroxide to form the more insoluble ferric hydroxide, so that less iron is maintained in solution. Continued corrosion can therefore be maintained by the aid of the dissolved oxygen, apart from its function in corrosion by differential aeration, so that effective removal of this gas from boiler feed is vital.

Although it is generally known that corrosion is combated by making the water alkaline, the reason is not always understood. The explanation is found in the difference between acid and alkaline waters from the viewpoint of hydrogen and hydroxyl ions. In an alkaline solution, according to the principles already outlined, there is an excess of hydroxyl ions over hydrogen ions. Such hydroxyl ions are potential partners of the dissolved ferrous ions to produce ferrous hydroxide as shown in the equation above. Since there is a balance at a given temperature between the two ions, with a constant concentration product, it follows that as the concentration of one type of ion increases that of the other must decrease. It follows that in an alkaline solution the increased number of hydroxyl ions results in the suppression of the

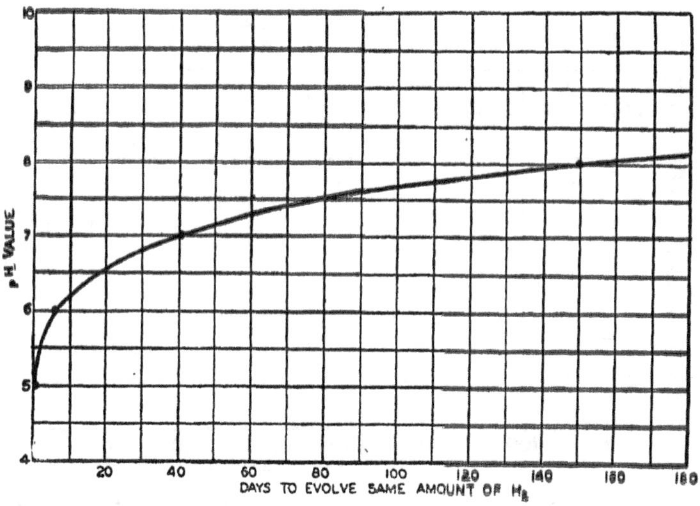

FIG. 25. Effect of Hydrogen Ion Concentration on the corrosion of iron.

solution of metallic iron and the diminution of corrosion. This is illustrated by the graph in Fig. 24, which shows experimental results[21] of the determination of the solubility of ferrous hydroxide in caustic soda solutions of various degrees of alkalinity; as the pH value increases—or more hydroxyl ions are present—the amount of iron in solution decreases, falling sharply after pH 9·5. This is confirmed by another investigation in which[22] it was found that the amount of hydrogen gas produced in the corrosion of iron was proportional to the hydrogen ion concentration up to pH 9·4, after which the evolution of gas almost ceases (Fig. 25).

These conditions, of course, are modified by other factors such as the presence of dissolved oxygen which can remove dissolved iron or hydrogen gas more quickly and thus enable the corroding action to continue. But it is clear why alkalinity reduces the aggressiveness of dissolved oxygen—the amount of iron in solution or the amount of hydrogen gas to be oxidized are reduced. In acid solutions the corroding reaction is fundamentally the same as outlined above, but in such conditions the hydrogen ion concentration is so greatly increased that corrosion is much more rapid; for instance, the more rapid hydrogen gas generation gives such increased pressure in the gas film that gas bubbles are formed which need no help from oxygen to escape.

High Temperature and Pressure Conditions.—In boiler plant practice the effect of temperature on hydrogen ion concentration is particularly important. As the temperature increases, water dissociates to an increas-

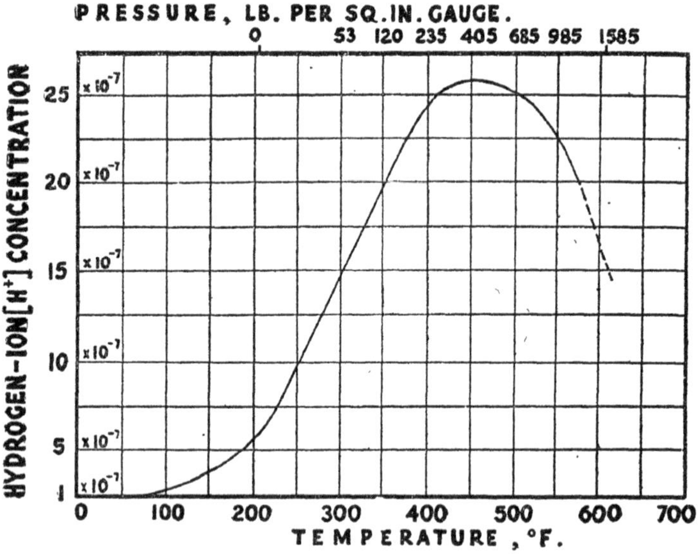

FIG. 26. Variation of Hydrogen Ion Concentration with temperature.

ing extent, the actual hydrogen ion concentration being shown[23] for various boiler temperatures and pressures in Fig. 26. From the standard unity (or pH 7·0) at 72° F. the concentration of hydrogen ions increases to as much as 25·8 times when the temperature is about 480° F. Returning to the definition of pH value, it will be recalled that pH 7·0 represents a concentration of 0·000,000,1 gram of hydrogen ions per litre of water or solution. At about 480° F., therefore, the concentration has increased to 0·000,002,58 gram per litre. On the pH scale this is between 0·000,001 gram per litre (pH 6·0) and 0·000,01 gram per litre (pH 5·0); actually, the hydrogen ion concentration at 480° F. corresponds to pH 5·6. In short, there is a corrosive concentration of hydrogen ions present under these operating temperature conditions.

This explains the corrosion which has been experienced on certain high-pressure plants,[24, 25] notably in connection with feed pumps, feed heaters, and economizers, although the pH value of the feed water has been very nearly 7·0 (and even a little over 7 in some cases). The corrosive power of a feed supply depends upon the hydrogen ion concentration under operating conditions rather than the pH value as measured on cold feed water. The reason for the lower limits imposed for the dissolved oxygen admitted in high-pressure boiler feed is also apparent from this potential increase in corrosive power. It is also evident that the feed to such a plant should have a pH value well above 7 to counteract this feature. Experience indicates that the pH value should be over 8·5 and preferably nearer 9. Ideally, for minimum corrosion a pH of about 9·5 is desirable, but other considerations such as feed-pump impeller materials, accumulation of alkalinity, etc., may not justify such procedure.

Effect of Saline Matter.—In general, it may be said that higher alkalinity is required to inhibit corrosion as the concentration of dissolved minerals increases. This aspect is illustrated by reference to Fig. 27, which shows the results of an investigation[26] of corrosion by

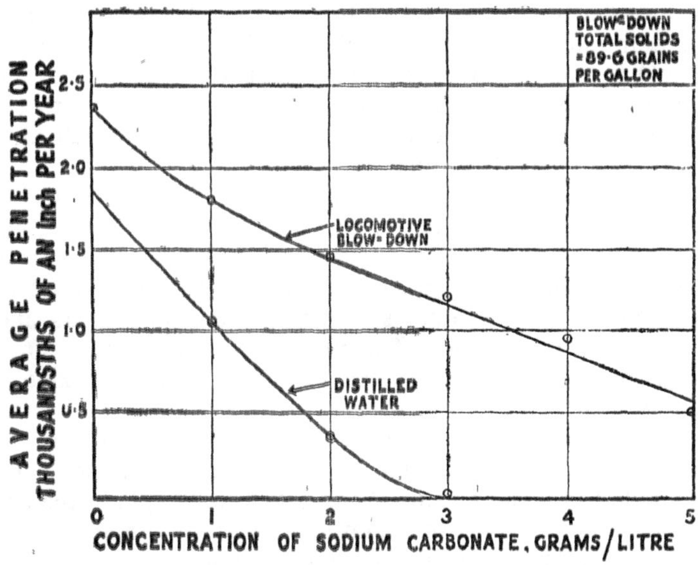

FIG. 27. Effect of saline matter on corrosion of iron.

distilled water and a blow-down liquor. Under the same alkalinity conditions, corrosion is considerably greater in the blow-down water. A similar state of affairs operates with dissolved oxygen; the saline matter increases the electrical conductivity of the water and there is a greater tendency to set up small electric cells owing to the differential aeration principle, resulting in localized pitting of the metal. The need for very low dissolved oxygen in high-pressure feed is thus again evident, to limit the alkalinity required for protection and therefore to limit the blow-down also. Similarly, the low total solid content desirable for high-pressure boilers serves not only to restrict blow-down, foaming and priming, etc., but is of indirect assistance in combating corrosion; as the dissolved solids increase, increased protective measures would be desirable.

Coagulation Control

The removal of suspended matter, colour, and bacteria from water is usually accomplished by the addition of alum, the alum combining with the natural alkalinity of the water to form aluminium hydroxide or other complex aluminium salts. Such compounds form as bulky precipitates which, in settling out in the settling tanks, carry the colour, suspended matter, and bacteria down with them. This operation is known as coagulation, and it takes place most completely and rapidly at a definite pH value. Owing to variation in colour, turbidity, natural alkalinity, source of supply, etc., this optimum pH value is different for different waters. Thus, Catlett[26] found that the optimum pH for waters in North Carolina varied from 4·3 for certain highly coloured waters, to 7·6 for some turbid waters which were free from colour. Again, Hatfield[27] found the maximum rate of flocculation with alum in Lake St. Clair water took place at pH 6·1 to 6·3, but the filter effluent was found to be free from soluble aluminium salts over a wider range, from pH 5·8 to pH 7·5. It is therefore necessary to determine the optimum for each individual water, and this can best be done by means of the "jar test."[28]

Jar tests are made as follows: Solutions of definite strength of alum, soda ash, and lime are made up, for example, so that 1 c.c. contains 1 grain of alum or alkali. Then take several jars of uniform size and shape and place in each 1 gal. of the water under test. First we add varying quantities of the alum solution to the different jars and note which gives the quickest floc. The pH of each jar is determined, and the one giving the best floc is taken as the optimum pH value for the coagulation of the water in question.

This is if the water contains sufficient natural alkalinity to give a floc. If it does not, a definite amount of alum is added to each jar and a varying amount of alkali added. Then, as before, the pH of the one with the best floc is taken as the optimum pH. After this, another set of tests should be made, decreasing the amount of alum; by continuing in this way the minimum amount of alum to be added can be found. This is advisable, since the same pH value may be obtained using smaller amounts of chemicals.

In this way the optimum pH value for a particular water is found, so that the minimum amounts of alum and alkali are used at all times, with the result that the best and quickest floc is obtained with no wastage of chemicals. Such control has resulted in reducing considerably the amount of chemicals used in a large number of water treatment plants; even more important is the fact that the water supply is more uniform in quality. Other advantages are that coagulation at the optimum means

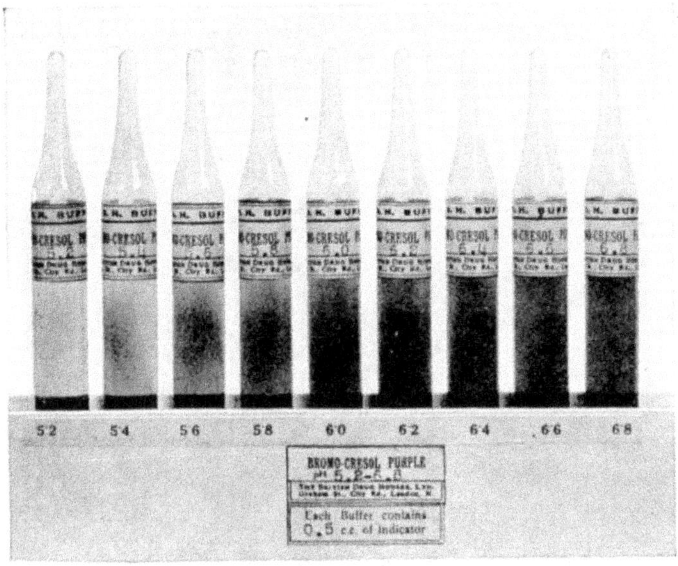

PLATE 18. SET OF STANDARDS FOR HYDROGEN ION MEASUREMENT

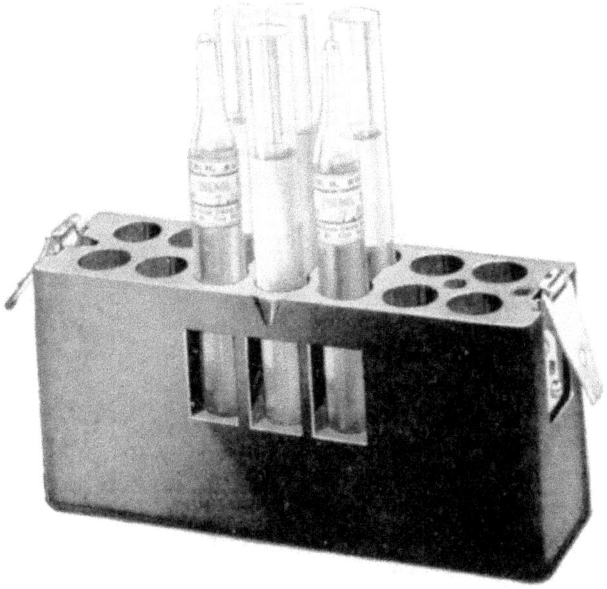

PLATE 19. COMPARATOR CASE IN USE

PLATE 20. NECKAR SODA AND SATURATED LIME-WATER SOFTENING PLANT

a better filter effluent, as it eliminates practically all solvent action on the floc and the appearance of aluminium in the effluent; it also increases the rate of filtration and makes possible a decrease in the size of the coagulating basin.

The use of alum, of course, involves an increase in the number of hydrogen ions, which tends to increase corrosion. In many cases it is therefore necessary to increase the pH (i.e., reduce the number of hydrogen ions) by treating with some alkali such as lime or soda either before or after filtration. By pH control the reaction of the water can be exactly regulated to the point where the smallest amount of corrosion will take place with the addition of the least amount of alkali.

Similar tests can, of course, be carried out with other coagulating agents, and the procedure can also be applied when de-oiling condensate with coagulants.

Softening Plant Control

Hydrogen ion control has been applied to lime-soda softening by Baylis,[28] who claims that more efficient softening can be obtained by

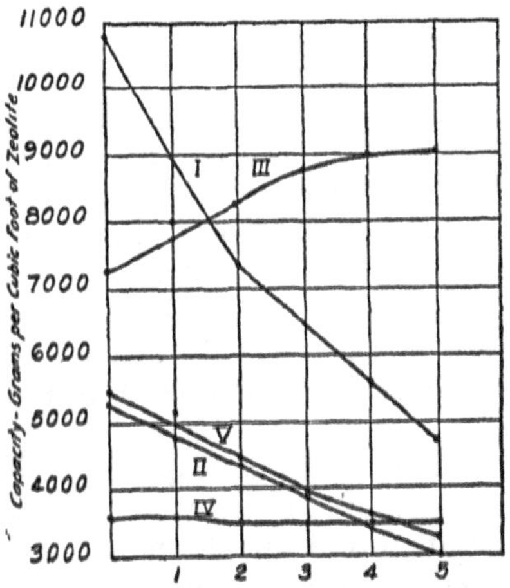

FIG. 28. Capacity of sodium zeolite on successive runs, using regenerating solution containing 0·1 per cent acid.

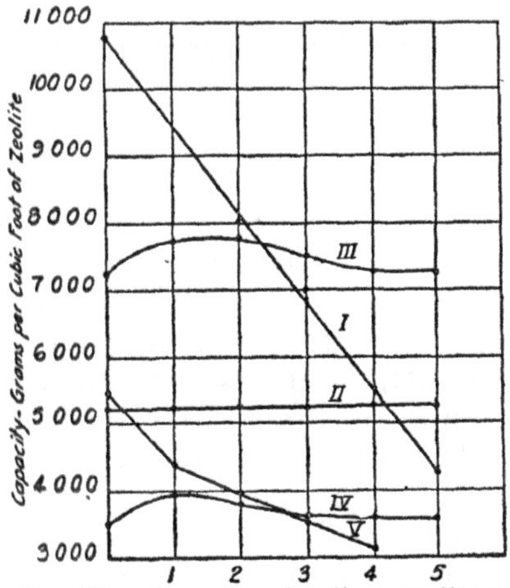

FIG. 29. Capacity of sodium zeolite on successive runs, using regenerating solution containing 0·1 per cent alkali. (Alkali is caustic soda.)

the use of pH measurements. It will be evident that conditions giving minimum solubility of calcium carbonate, etc., are of great importance in water softening. Baylis found that calcium carbonate reached its minimum solubility at pH about 9·4, while for magnesium hydroxide the pH for minimum solubility is 10·6. These values have also been determined by Greenfield and Buswell,[29] who found that precipitation of magnesium as magnesium hydroxide is complete at a pH of about 10·6, and that the precipitation of calcium carbonate is complete at pH 9·5.

Measurements of hydrogen ion concentration can therefore be used as a guide in the process of water softening. Chemists in water softening plants are aware that the treatment they are giving the water is producing approximately a certain pH, but this is not their guide for applying the chemicals. While concentration measurements can be used with fairly good results, greater economy and more efficient softening will result from application of pH control as well.

Base-exchange softening is also affected. Sweeney and Riley[30] have shown that one of the factors profoundly affecting the functioning of the softener is the hydrogen ion concentration prevailing in the water to be softened or in the solution used in the regeneration. The capacity, continued activity, and general performance of the plant are seriously affected by this factor. An indication of the magnitude of its influence on a selection of commercial zeolites will be obtained by reference to Figs. 28 and 29, which show the marked effect on capacity, but the original paper should be consulted for full details.

The importance of hydrogen ion concentration in preventing corrosion and deposition in recirculating cooling systems also is emphasized in research by Thompson and Ryznar.[31]

Boiler Water Recirculation

Corrosion due to increased hydrogen ion concentration at high temperatures and pressures has been described above. The modern remedy is the comparatively simple procedure of recirculating a small proportion of the boiler water through feed-pumps and other affected auxiliary units. The increased alkalinity so imparted to the feed supply increases the protective hydroxyl ion concentration. Although basically a simple operation, recirculation requires careful consideration to obtain full advantage from its application, particularly as it is *not* invariably suitable for high-pressure plant operation.

Recirculation is superior to direct addition of alkali reagents to the feed for this purpose, since no additional reagents are added to the water, and thus there is no rapid concentration of alkalinity in the boiler water. Blow-down conditions therefore remain normal, and furthermore the amount of boiler water recirculated is readily controlled by suitable orifices, etc. Recirculation should not be confused with blow-down. The mineral content of the system remains unchanged, since the salines withdrawn from the boiler in a concentrated form are diluted by mixture with the ingoing feed water and are finally returned to their source—the boilers. The solids so introduced into the feed water are thus to be considered distinct from the solids originally present in the feed supply. The solids present in the original feed will concentrate in the boiler in the normal manner and must be controlled by normal blow-down practice. The solids in the recirculated boiler water are merely circulating, and are not lost from the system.

In addition to the improved alkalinity and hydrogen ion concentration conditions, other advantages are gained in recirculation. A common method of dealing with traces of dissolved oxygen which may pass the de-aerator is the continuous addition of small amounts of sodium sulphite to the feed. To ensure rapid reaction, it is advisable to have at least twice the equivalent amount of sulphite in the water, and it follows that if this were added directly to the feed increased blow-down would be necessary to handle the increased solids entering the boilers. By recirculation, excess sodium sulphite from the boiler water is introduced to the feed prior to the economizer and economy in treatment is achieved since this additional sulphite is available for reaction with the dissolved oxygen to protect the economizer. Furthermore, the alkalinity also imparted to the feed reduces the aggressiveness of any traces of dissolved oxygen present. Reliable continuous protection of the economizer is thus secured. It is also possible to add chemicals to individual boilers via the recirculated water, thus permitting variation in conditioning treatments for individual boilers if required by operating conditions. Steam binding in the economizer during varying load periods is also said to be eliminated due to the washing effect of the recirculated water.

The application of recirculation procedure is limited by the condition of the feed water. If the feed consists very largely of condensate, it is probable that this method will be a successful and economical means of protecting economizers, etc. But if the feed contains a good deal of make-up it is likely that such a large percentage of boiler water will have to be recirculated that the process will not be economical or practical. The explanation is found in buffer action, explained on page 118. Feed water consisting principally of condensate is an unbuffered solution, and the pH value will change very largely with small additions of alkalies such as are contained in the recirculated boiler water. On the other hand, raw feed or feed containing a large proportion of treated make-up, etc., contains sufficient minerals to set up buffer action, and substantial amounts of alkali will then have to be added to effect changes in the hydrogen ion concentration. Occasionally, in such cases, conditions may be such that the recirculation of a small amount of boiler water may be supplemented by the continuous addition of a strong caustic soda solution to the feed. It is clear, however, that such a system would need very careful control of resulting alkalinity.

It has already been shown that the condition of minimum attack of iron is represented by a pH of 9·6. Operating troubles are likely with pump impellers if the alkalinity is increased to this point, and in recirculation practice it is sufficient to limit the pH to 8·5–9·0, since this will reduce corrosive attack to a minute proportion of that which would occur if the pH remained at about 7·0 for high-pressure feed. At one power plant,[25] however, a special recirculator has been designed to pass the water from the drum of the 1,300 lb. boilers direct to the economizer inlet, and in this case the pH has been taken over 9·0.

Some existing installations[24, 25] afford excellent illustrations of the careful study of plant conditions which precede successful operation.

The location of the boiler water take-off is important. The water may be collected by means of a header extending the length of the boiler drum and perforated with holes throughout its length ; these holes, however, should not face any streams of water, etc., emerging from the circulating tubes of the boiler. Selection of a suitable take-off point should be based on the water analysis, so that the most concentrated and least cloudy water can be recirculated at all times. The presence of much suspended matter might endanger recirculation by clogging orifices, etc., and it is clear that the greater the concentration of the water the greater its effect and therefore the smaller the volume required for recirculation. In some instances the water has been introduced at the heater storage tank, so that its mixture there can provide protection for all equipment between the de-aerating heater and the boiler. If the boiler water were mixed with feed prior to de-aeration, any sulphite in the recirculant would be consumed instead of being available for reaction with the last traces of dissolved oxygen left in the feed after de-aeration, as is preferable. An alternative procedure is to recirculate the boiler drum water to the feed water cycle at the economizer inlet. This has the advantage that the power required is only that sufficient to overcome economizer friction ; the power involved is about 6 h.p. for a 150,000 kw. station. The power involved in the case of circulation to the heater storage tank is indicated by the instance of a 55,000 kw. station which required 40 h.p. pump capacity to give protection of this nature.

The complete plant cycle must be carefully considered in relation to recirculation with a view to estimating any effects on the functioning of various equipment, heat balance, etc. For instance, at a plant operating on feed which is principally condensate, the introduction of recirculation could cause difficulties if the feed water were supplied to de-superheaters, and after recirculation it might be necessary to make special arrangements for continuing the supply of condensate only to such de-superheaters. The effects of recirculated salines on any chemical treatment introduced into the feed system as a conditioning process also would require study.

Operating results for a number of stations employing recirculation in U.S.A. are summarized in the following table:

TABLE 12

Boiler Water Recirculation Data

	Feed Water				Boiler Water		Operating Pressure	
	Original		After recirc.					
Plan No.	Total solids p.p.m.	pH	Total solids p.p.m.	pH	Total solids p.p.m.	pH	Lbs. per sq. in.	Amount of boiler water recirculated
1.	7·0	6·7	16·0	8·8	867	10·9	650	0·5
2.	1·4	7·5	25	9·2	1,692	11·0	1,300	5·0
3.	27·6	7·3	48	8·6	1,000	11·2	400	1·0
4.	7·3	6·9	14	8·4	1,200	11·1	650	less than 0·5
5.	1·4	7·5	37	9·4	1,728	11·0	1,300	5·0
6.	4·5	7·0	14	8·6	573	11·1	650	1·0

Although heavy feed-pump maintenance, etc., was experienced originally in some of the above stations, or corrosive attack was noted in service by checking the iron content of the water, the adoption of recirculation has been completely successful in obviating pump troubles, economizer corrosion, etc. In the case of the 400-lb. plant mentioned, recirculation has been employed since 1935, with complete success. Experience with a plant subject to appreciable load variation throughout the day is of considerable interest. The boilers operated at 410 lb. pressure and it was found before recirculation that the greatest tendency to produce low pH condensate occurred at low load periods. After adopting recirculation it was found that pH values could be maintained fairly steady around the desired average with comparatively little attention throughout load fluctuations. This recirculation of less than 1 per cent of boiler water thus gave full protection to the plant even at times when formerly corrosive hazards were aggravated. Experience has also shown that sharp reductions in load have not caused increased solids in steam, and consequent danger of superheater deposition, etc. It might be expected that a decrease in load would result in a sharp increase in the saline content of the feed, leading to risk of priming, etc.

Finally, it should be pointed out that recirculation will in effect reduce blow-down since a lower concentration can be carried in the feed water. With recirculation, protective alkalies are introduced in the feed cycle to ensure full protection and consequently less minerals need be passed to the boilers in the original feed water. The lower the feed water concentration, obviously, the lower the blow-down, and the less evaporator distillate which must be provided to replace the blow-down. The importance of this feature thus varies according to local plant conditions and the extent to which condensate is fed to the boilers.

Electrical Methods of Preventing Corrosion

The fact that the modern explanation of corrosion is electrochemical in nature has led to many attempts being made to combat the trouble by electrical methods. These do not therefore depend on the removal of the corrosive gas or gases but on the superimposition of polarity. Since corrosion occurs through the medium of numerous minute electric cells, it should be stopped by setting up a stronger cell of which the cathodeplates are the plates of the boiler shell. The boiler heating surface cannot then become anodic and attackable. This is accomplished by placing a small metal anode inside the boiler and applying a small continuous current. The water is then the electrolyte and instead of the metal of the boiler being attacked, dissolution of the anode takes place slowly and continuously. Cases have been described where this method of protection has been successfully applied to water tube boiler and locomotive boilers. Some electrical treatments are described in the section of this book dealing with scale prevention. Some of these impose polarity on the boiler shell, and thus combat corrosion.

The use of zinc plates or blocks is sometimes adopted to prevent corrosion and the principle underlying such procedure is the same as that of the above electrochemical method of protection. Attack is concentrated on the zinc, which is cheap and easily replaceable. The purity of the zinc is a big factor in the effectiveness of this expedient. Chemically pure zinc is too resistant for the purpose and it is therefore usual to give a specification of the maximum impurities to be present. It is usual to ask for at least 98 per cent zinc and the iron content must be kept low, usually of the order of 0·05 per cent. Zinc rods and blocks are of little use in condensers owing to the small difference in potential between the brass of the tubes and the zinc.

REFERENCES

(1) D. K. French, " Ind. Eng. Chem.," 1934, p. 1319.
(2) See Nat. Elect. Light. Assoc. Report for 1930.
(3) C. E. Joos, " Mech. Engineering," March 1938, p. 229.
(4) Richter, " Chem. Zeitung," 1933, p. 241.
(5) Zapffe, Trans. Am. Soc. Mech. Eng., Feb. 1944.
(6) Schroeder & Berk, " Combustion," Aug. 1940, p. 19.
(7) Colbeck, Smith & Powell, Proc. Inst. Mech. Eng., Nov. 1942.
(8) C. W. Rice, " Combustion," Jan. 1941, p. 37.
(9) Zapffe, Trans. Am. Soc. Mech. Eng., Dec. 1942.
(10) Illinois Univ. Eng. Station Bulletin No. 177, 1928.
(11) Splittgerber, " Gesundheitsingenieur," 1927, p. 811.
(12) Berl, " Forschungsarbeiten auf dem Gebiete des Ingenieurewessens Heft," 1927, p. 295.
(13) Am. Soc. Mech. Eng., Symposium on Caustic Embrittlement, " Combustion," Dec. 1941, p. 35.
(14) Schroeder, Berk & Partridge, " Proc. Am. Soc. Test. Mat.," 1936, p. 721; Straub & Bradbury, " Trans. Am. Soc. Mech. Eng.," 1936, p. 589.
(15) Straub & Bradbury, " Mech. Engineering," 1938, p. 371.
(16) J. B. Romer, " Mech. Engineering," Feb. 1939, p. 145.
(17) U.S. Patent 1,910,413.
(18) List, " Vom Wasser," 1932, p. 168; Splittgerber, " Vom Wasser," 1931, p. 93.
(19) Purcell & Whirl, Trans. Am. Chem. Soc., 1943, p. 343–359.
Purcell & Whirl, " Mechanical World," Dec. 31st, 1943, p. 757.
(20) J. H. Walker, " Mech. Engineering," 1942, p. 891.
(21) Whitman, Russell & Davis, J. Am. Chem. Soc., 1925, p. 70–79.
(22) Shipley & McHaffie, " Can. Chem. Met.," 1924, p. 121.
(23) Based on International Critical Tables.
(24) Sperry, " Combustion," June 1939.
(25) Baker, " Combustion," Oct. 1940, p. 31.
(26) Catlett, J. Am. Waterworks Assoc., 1924, p. 887.
(27) Hatfield, ibid., p. 554.
(28) Baylis, " Ind. Eng. Chem.," 1928, p. 1191.
(29) Greenfield & Buswell, J. Am. Chem. Soc., 1922, p. 1435.
(30) Sweeney & Riley, " Ind. Eng. Chem.," 1926, p. 1214.
(31) Thompson & Ryznar, " Combustion," Nov. 1942, p. 41.

PART FOUR

FOAMING AND PRIMING

WHEN WATER IS CARRIED OVER FROM THE BOILERS WITH THE STEAM the two terms "priming" and "foaming" are frequently used indiscriminately to describe the process. These terms are not really interchangeable, and in view of the confusion in the use of terms and their meaning it will be convenient first of all to obtain a clear idea of what is meant by foaming and priming.

Several types of behaviour can be distinguished. Firstly, there is rapid or too violent boiling, with consequent mechanical projection of droplets or even slugs of water into the steam pipe. This more or less violent bursting of steam bubbles is "priming." On the other hand, if the conditions are such that the bursting of the steam bubbles is delayed, then a foam of bubbles of steam will form on the surface of the water. According to the stability of these bubbles, this foam may range in thickness from a skin or layer barely covering the surface to a mass that fills the steam space entirely. This may properly be described as "foaming." Finally, there is the behaviour analogous to "bumping" in the laboratory beaker or kitchen kettle. In such cases the formation of steam bubbles is delayed and violent ebullition occurs intermittently. The water remains calm until it is sufficiently superheated to cause sudden formation of large steam bubbles which burst violently at the surface and throw water over into the steam pipes. These phenomena are probably closely related, and it will often be difficult to distinguish between cases in actual practice, the carry-over of water in many cases probably being due to a combination of these processes. In discussing causes of priming and foaming it will not be necessary, therefore, to consider these separately, and for convenience the terming "priming" will be adopted to describe all these conditions.

It should be noted that the situations described above are not necessarily descriptions based upon actual observations of steam boilers in operation, but are to be considered more as suppositions based upon observation of similar behaviour in glass flasks or open metal vessels. In fact, the only record of an actual observation in a steam boiler which the author has been able to find is that of Joseph and Hancock.[1] This is so interesting that it may be described in some detail. Their observations were made with a small laboratory boiler working at 100 to 150 lb. pressure, the boiler being specially fitted with sight glasses. They were thus able to observe the water surface and the formation and the breaking of the bubbles, and, also, by certain adjustments, a view through the water itself could be obtained. They found that with the steam valve shut steam bubbles were formed at the heating surface and passed through the liquid, always breaking at the water surface. With all waters, even distilled water, whenever the steam valve was opened the water space became filled with innumerable bubbles, some of which did not rise rapidly to the surface, but remained almost suspended in the water. Further, when the valve was open the bubbles rising to the steam space did not break immediately, but formed a layer upon the water surface. This layer grew in thickness as the valve opened wider, and it rose higher and higher into the steam space until finally it reached the valve, when priming followed. With distilled water no such carry-over of water occurred. There was a thin layer of bubbles on the surface of the water, but these broke too rapidly for wet steam to result. On the other hand, "when a natural or artificial saline water is evaporated the concentration, of course, steadily rises as evaporation proceeds, and when the steam valve is closed the appearance of the water surface in no way differs from that of distilled water, even when the concentration of sodium salts is as high as 8,000 parts per million. But as evaporation proceeds and concentration increases, the bubbles gradually increase in size, number and stability, and rise more and more into the steam space until they ultimately fill it—no appearance of priming is seen while the valve is shut, and with any given solution it has to be opened to a definite extent before priming can be observed. . . ." But more of this later.

When the foam fills the steam space entirely the liquid films around the bubbles of steam are drawn into the steam pipe and wet steam results. There is also the further disadvantage of wastage of fuel, since the hot boiler water carried over in this way does no useful work. Further the salts in the water may cause damage to steam pipes, valves, cylinders, etc., while the lubricating value of the oil in the cylinder may be destroyed, with consequent overheating. A case of a battery of five boilers has been quoted, each evaporating 20,000 lb. continuously. At the rate of 0·04 lb. of impurities per 1,000 lb. of steam, over 25 cwt. of impurities would be carried over in a month's continuous operation. The seriousness of such a situation scarcely needs emphasis, but consideration of the scouring effect of that quantity of dust (after passage through superheaters) on steam plant valves, turbines, etc., will point to the desirability of eliminating foaming and priming as completely as possible. Finally, in some cases priming may be so bad as to cause the gauge glass to become filled with bubbles, thus making it difficult to determine with accuracy the level of the water in the boiler.

Causes of Priming

The causes of priming may be divided into two groups, physical and chemical. Among the physical causes the design and operation of the boiler are the most important. The more complex the interior—that is, the more of a maze of tubes, stay-bolts, etc., there is in a boiler, the more liable is this operating trouble to occur. Then, again, the smaller the steam space above the water the more likely is water to be carried out. This is only natural, since the layer of foam need not be so thick to reach the valve. The work of Joseph and Hancock already referred to is convincing proof that the lower the water level the less priming will there be. For example, a 2½ in. drop in the water level in one test reduced the priming from 90 per cent to nil.

There are several ways by which violent ebullition can occur. The loosening of a piece of scale would bring water into contact with overheated metal and violent boiling would occur for a few seconds. The same effect would be obtained by the sudden introduction of solid matter, whether it entered with the feed water or was produced by the loosening of scale. Another way of assisting rapid boiling is by a rapid fall in pressure, such as may be caused by the sudden opening of a valve. Generally speaking, the faster steam is taken from a boiler the more readily will the boiler prime. These statements are confirmed by the observations of Joseph and Hancock. They found that if the pressure is increased, the priming is greatly reduced. It is also evident from the account of visible observations recorded above that the faster steam is taken from the boiler the more priming obtained. The steam valve had to be opened to a definite extent with any given solution before priming could be observed. In other words, priming depends largely on the rate of evaporation. They also record some experiments with solutions of sodium chloride, sodium sulphate, and sodium carbonate; here also an increased rate of evaporation led to increased priming, the sodium sulphate being the worst offender. This work therefore confirms the fact that boilers with a low rate of evaporation per sq. ft. of heating surface are less liable to priming than those in which the rate of evaporation is higher. Hence the greater freedom of the Lancashire-type boilers compared with the multi-tube Stirling, Babcock and Wilcox, and Yarrow types. Other physical causes are uneven steaming of a boiler due to patchy scale, absence of baffle plates in steam drums, vibration and rolling of boilers, etc.

When we come to consider the chemical causes we find that various substances are believed to cause priming. Thus "organic matter," "oily matter," "alkalis," "soap and soapy substances," "saline matter," and "suspended solids," to quote a few of the terms widely used, have been blamed for this trouble. There is also a general belief that sodium salts are the mischief makers; in fact, these are the only definite chemical substances to which priming has been attributed, although there seems to be very little experimental evidence that sodium salts alone are responsible. On the other hand, there are many contradictions to this notion. For instance, Lord Rayleigh[2] stated that he had found that even a strong salt solution (sodium chloride) did not foam much. Tatlock and Thompson[3] cite a case in which a boiler water contained 16,000 parts per million sodium carbonate, 2,100 parts per million caustic soda, and 13,000 p.p.m. sodium sulphate, but showed no trace of priming. Koyl[4] also describes instances in which locomotive boiler water contained high concentrations of sodium salts, but no priming occurred unless finely divided solid matter was also present. Laboratory tests by Foulk[5] also showed that solutions of the chloride, sulphate, hydroxide, carbonate and bicarbonate of sodium, and solutions of various mixtures of these salts gave little evidence of priming, even when the solutions were taken almost to saturation. On the other hand, McGill[6] found that foaming or priming took place at 10 lb. pressure when the concentration of sodium salts reached 1,500 parts per million expressed as sodium oxide, and he presumes the effect would have been much increased in a boiler working at 200 lb. pressure. Also, Joseph and Hancock[1] found that sodium chloride, sodium carbonate, and sodium sulphate caused priming, though the usual opinion that the carbonate is especially objectionable was not confirmed by their observations. In these tests the sulphate was the worst at the lower concentrations, though at concentrations of the order of 4,000 p.p.m. there was not much difference. These tests also were carried out with actual boilers, each test being carried out in three stages; first, for small scale laboratory tests an experimental boiler working at 100 to 150 lb. pressure was used; the second stage was carried out in a stationary boiler of the locomotive type, having a capacity of 600 gal.; and the third in locomotives in ordinary service. The argument has been used that the evidence against sodium salts has been obtained with actual boilers under pressure, whether laboratory experimental units or actual working boilers, while the proof that sodium salts alone do not cause priming has been obtained from experiments in flasks. As will be seen even from the above few references, this is not altogether true.

Some technicians maintain that finely divided solid matter—"suspended solids"—is the responsible agent. This also is subject to objections, in fact most of the evidence disproves it. Thus, Joseph and Hancock found that powdered calcium carbonate alone, even to the extent of 4,000 p.p.m., caused no priming; McGill observed no priming with water and powdered chalk; Foulk tried various kinds of insoluble powders, including boiler scale, but obtained no appreciable foaming. Some observations recently published by Behrman[7] describe the effect of organic matter and also suspended solid matter on a foaming boiler water. The sample of water used was taken from a boiler which foamed so badly that the highest concentration of dissolved solids which could be carried under normal operating conditions was about 2,250 parts per million. The boilers were of Stirling type, provided with excellent circulation and adequate steam space. The boiler blow-down contained, amongst other things, 2,600 parts per million dissolved solids and 200 p.p.m. suspended matter, while its colour was an intense dark brown. Removal of the suspended matter and some of the colloidal organic matter was effected by coagulation and filtration. Foaming was greatly reduced. In these tests some of the colloidal organic matter had not been removed, so steps were taken to do so by other means. Foaming was still further reduced. Another example is cited by Hancock,[8] the case of a river water with 1,000 p.p.m. of solid matter, of which 25 p.p.m. were organic matter. This water foamed badly, but after removing the organic matter by treatment with charcoal foaming was prevented.

Whatever may have been considered to exert the predominating effect in promotion of foam and subsequent priming, most workers in this field have been of the opinion that it was essential to have soluble salts and/or solid suspended matter present to cause foaming. Other substances and conditions may by secondary influences assist in the production of foam, but it has been generally recognized that these two classes of matter are the chief responsible agents in a large number of cases. The whole problem has been greatly clarified in recent years by important investigations on the foaming of boiler water carried out by Foulk in the United States of America. The results of these researches reveal why it has been so difficult in the past to allocate the blame for foaming and priming.

Balanced Layer Film Formation

Fundamentally, this title of the present-day explanation of foam formation simply means that foaming is supposed to be set up by a difference in concentrations at parts of the boiler water or other solution. In the body of a solution, the molecular and ionic forces are free to balance themselves. At the surface, on the other hand, the forces are balanced by corresponding forces only on the liquid side. There are therefore unbalanced forces corresponding to the gas side of the surface, and such unbalanced stresses give rise to the well-known surface properties observed in all solutions. Now the result of this situation is that the solute (dissolved substance) tends to concentrate in the body of the liquid or in the surface layer of the solution. This means there will be a difference in concentration of solute in the surface layer as compared with that in the bulk of the solution. Whether the surface layer will be the more concentrated or otherwise depends upon the physical and chemical properties of the dissolved substance. This, however, is immaterial for our purpose. The same effect will occur at any gas-liquid interface (such as a steam bubble) in the solution. Thus, with each steam bubble there will be an interface between the gas and the bulk of the solution or boiler water. The concentration at the surface layer, i.e., inside the bubble and next to the gaseous steam, will differ from the concentration on the outside of the bubble which is the solution. In short, there is a film around the bubble, and the concentration in this film differs from the concentration in the body of the solution or boiler water. It may be greater and it may be less than the concentration of the body of the solution. When the substance concentrates in the interface or film it is said to be "positively adsorbed," whilst if it tends to concentrate in the body of the solution it is said to be "negatively adsorbed." When the bubble of steam is emerging through the surface of the boiler water, the solution will drain away from around the bubble, leaving a film of solution which will be a stable film or an unstable film.

If the steam bubble is to burst when it reaches the surface of the boiler water, it is evident that the water produced by breakage of the film will have a different concentration from the surrounding boiler water or solution. It cannot remain so ; the concentration must become uniform ; by the laws of thermodynamics, therefore, some work must be done by some agency to restore the equilibrium of uniform concentration. What this means is that the system including the steam bubble will tend to resist forces which attempt to rupture the bubble. In short, there will be a tendency to maintain the bubble, and many such bubbles will form a foam. There will be a " balanced layer " of solution around each bubble.

Naturally, the continued existence of a bubble depends upon how strongly it is able to resist the forces tending to equalize the concentrations, or break down the film or balanced layer. For example, if the concentration difference is great, the bubble may be able to resist rupture ; thus pointing to the undesirable character of high saline concentrations in the boiler water. But before considering this fully, it may be useful to point out that this balanced layer idea can be confirmed experimentally.

If the theory outlined above is true in principle, it should be possible to mix substances, say one which concentrates in the body of the solution (or is negatively adsorbed) and one which concentrates in the interface (or is positively adsorbed), so that the concentration is equal throughout the solution. Such a solution should not foam. This has actually been done in practice. Soap and sodium sulphocyanate both concentrate in the interface. Sodium chloride and sodium sulphate concentrate in the body of the solution. Mixtures in the right proportions of (*a*) soap and sodium sulphate in water, (*b*) sodium sulphocyanate and sodium sulphate, and (*c*) sodium sulphocyanate and sodium chloride gave mixed solutions of non-foaming properties. It is evident from Foulk's balanced layer explanation, too, why a pure liquid does not foam ; there can be no adsorption or establishment of differing concentrations.

Foam Stabilization

The above explanation developed by Foulk is a reasonably good mechanism for foam formation, but one is immediately faced with the question as to why some foams are more stable than others. This introduces the question of foam stabilization, but the means by which some foams become sufficiently stable to cause priming troubles is an incompletely solved problem. It is known that foams may be stabilized by the addition of some substances to the solution, but while foams are stabilized at some times, the establishment of apparently similar conditions another time does not produce the stabilizing effect. Extensive studies of these factors have been made by Foulk and his collaborators over some fourteen years, and a detailed summary of his observations over that period has recently been published.[22] Readers are referred to that publication for details, but the main points of investigations are briefly reviewed in the following discussion.

In the first place the chemical conditions necessary for the formation of a foam may be described. A quotation[9] may be given on this point: "To get a foam the only essential is that there shall be a distinct surface film, in other words, that the concentration in the surface layer shall differ perceptibly from that in the mass of the liquid. All true solutions will, therefore, foam if there is a marked change of surface tension with concentration, regardless of whether the surface tension increases or decreases. . . . To get a fairly permanent foam, the surface film must either be sufficiently viscous in itself or must be stabilized in some way. This can be done by introducing a solid powder into the interface. Solutions of aqueous alcohol, sodium chloride, and sulphuric acid all foam when shaken, but the foam is unstable. Soap solutions foam when shaken and the foam is, or may be, quite stable, owing to the viscosity of the soap films. With saponin the surface film is even more stable. If we add to aqueous alcohol some substance like lycopodium powder, which goes into the interface, we get a stable foam. The same thing can be done with aqueous acetic acid by adding lampblack. The presence of enough finely divided solid in the interface will make the film so viscous that the foam will be quite stable."

The general idea at present, based on Foulk's and other work, is that suspended solids may lead to priming through stabilization of the foam, but it is by no means invariably true. The behaviour of suspended matter is so uncertain that it is evident the mechanism of foam stabilization (at least under boiler and similar conditions) is not a simple admission of solid particles to the interface. Consideration of some of the main observations will emphasize this point.

Consider first the work of Foulk himself. It has already been stated that he obtained no priming with either sodium salts or suspended matter alone. He also found that, when present in sufficient concentration, any one of the sodium salts together with any one of a variety of insoluble materials produced a stable foam. If this explanation is correct almost any soluble substance, along with appropriate solid matter, should produce foam. Accordingly, Foulk ran a series of tests with miscellaneous soluble material and powdered boiler scale or limestone as the solids. He sums up the results as follows: Calcium chloride, potassium chloride, hydrochloric acid, acetic acid, potassium dichromate, sugar, alcohol, and acetone in water solution along with finely divided solid matter, all gave priming mixtures. None of these substances produced stable foam in water alone without the presence of solid matter. Finally, tests were made at various pressures up to 50 lb. No effect was observed. The observation by Koyl that both sodium salts and finely divided solid matter was necessary for priming is noted above. Another interesting case mentioned by him deserves mention: On one part of a railroad the locomotive boilers did not prime, though there was scaling, due to the character of the water. When, however, these boilers were filled with softened water at a certain station they began to foam. This was at first blamed on the sodium salts in the softened water, but it was eventually proved to be finely divided scale which was loosened by the new water. This was demonstrated by using the treated water exclusively in several runs with a clean boiler, when no priming was obtained. A similar case has been described by Powell.[10] In this case a boiler commenced foaming badly at a change-over to zeolite-softened water and it became evident that the scale had been loosened by the softened water, since priming ceased when the scale was removed. These cases are what would be anticipated from Foulk's theory of the mechanism of foaming. The theory also accounts for the observations of Lord Rayleigh and of Tatlock and Thompson. On the other side we have the work of McGill and of Joseph and Hancock. McGill found that sodium salts alone gave priming, if present to the extent of 1,000 p.p.m., but he also records the fact that the presence of solid particles invariably increased the violence of the priming. Joseph and Hancock tested waters containing sodium chloride and calcium carbonate, and others in which calcium carbonate was formed in situ by interaction of sodium carbonate and calcium chloride, and found that priming was not increased by the presence of 2,000 p.p.m. of calcium carbonate. When they used powdered boiler scale and a water containing 4,000 p.p.m. of sodium chloride, the priming with the scale was actually less than without it. This led Foulk to carry out further tests on the effect of suspended solids on foam stability. He used pumice, sulphur, pyrolusite, bone black, boiler scale, etc., and found that the foam stabilizing effect was specific to the nature of the substance. Among the substances tried, powdered boiler scale and limestone were the most effective as foam producers under atmospheric pressure conditions. Decrease in particle size produced an increase in foaming. It was also found that increase in the amount of suspended solids had the same effect as an increase in the concentration of soluble salts, one being able to replace the other. This observation is in agreement with a similar conclusion of Beale and Stevens which will be referred to below. In fact, according to Foulk's tests it appeared that the solids which stabilize the foam are those which are not readily wet by the solutions and therefore float in a finely divided state in the interface of the steam bubbles. Thus, finely divided flint is instantly wet by water and so sinks to the bottom and does no foam stabilizing. Adsorbed oil on the surface of solids also seemed to be a big help in promoting foaming, and it was found that oil-free boiler scale lost much of its foam stabilizing power. Again, it was found that the foam stabilizing properties of solids might be lost by prolonged boiling, the time of boiling diminishing as the pressure increases. This was probably due to modification of the surface of the solid or some other change by which the wettability was increased. The observations of Joseph and Hancock are thus not necessarily rigidly true for all cases. They may have had readily wetted scale, etc., and from Foulk's observations it is fairly certain that calcium carbonate generated in situ (as in some of J. and H.'s tests) in the solution would have little foam stabilizing properties, since it would already be very thoroughly wetted.

It has also been found that calcium carbonate produced inside a boiler by decomposition of bicarbonate decreased priming, although when calcium chloride was pumped into a solution of sodium carbonate (producing calcium carbonate as a precipitate) the precipitate increased priming. In another case, magnesium hydroxide (formed by pumping magnesium chloride into caustic soda in a boiler) reduced priming and in fact tended to counteract the effect of calcium carbonate when present in sufficient amount.

Further research may eventually disclose the reason for the variable effects of solid particles. A more recent suggestion by Foulk is that electric charges on the bubbles and the solid particles may be determining factors, presumably in controlling the admission of solid particles to the interface and consequent stability of the foam. For the present, however, boiler operators can safely adopt the idea that foams need to be stabilized before they cause priming; and that while suspended solids and such matter may not always function in that direction, it is a good rule to maintain their concentration at a minimum. Oil and grease concentrate in the interface and are known to cause stable films in most cases.

Prevention of Priming

The physical methods may be considered first. In the design of boiler, the water level should be kept as low as possible consistent with safety for the particular boiler and working conditions of the plánt in question. Sudden opening of large steam valves, overloading, or undue forcing should also be avoided.

Among the chemical methods of prevention, complete exclusion of oil or grease should be aimed at. Oils are particularly dangerous when they contain saponifiable matter, as in the case of blended cylinder oils. This saponifiable matter becomes converted into a soap in an alkaline boiler water and very small amounts of such a soap will cause pronounced foaming and priming.

As regards saline concentration and the amount of suspended solids, the greater the quantity of dissolved solids the smaller the amount of suspended matter which should be allowed. Beale and Stevens,[11] for instance, have recommended the following operating limits based on figures obtained during several years from a large number of boiler installations:

Suspended solids	1	5	10	15	20	30	40
Total solids .. .. .. (Suspended plus dissolved)	2,500	1,600	1,100	700	550	400	300

These figures are expressed in grains per gallon and are comparable with similar figures based on approximately 175 per cent rating for land water-tube boilers which have been published in U.S.A.:

Suspended solids	1	6	10	15	26	50
Total solids .. .. (Suspended plus dissolved)	2,000	1,000	600	400	200	100

These figures also are in grains per gallon. The permissible amount of dissolved solids therefore diminishes rapidly as the suspended matter increases. While this effect will vary from plant to plant, according to the influence of boiler design and operation, it is a vital factor in the prevention of foaming and priming. The suspended solids should therefore be kept as low as possible. In this connection it may be recalled that sodium aluminate treatment has already been mentioned (under scale formation) for good coagulation of precipitates and suspended solids and has been found to reduce foaming trouble considerably. This chemical is also useful for keeping down the alkalinity of boiler water (for reasons already described), which is another important factor in preventing priming. It is generally recognized that the dissolved solids should be kept below a certain figure, varying according to individual plant conditions; as a general rule, however, it is accepted that the concentration of salts should not exceed 1,000 grains per gallon for boilers of the Lancashire type, while half this figure is taken for water-tube boilers. Much, however, depends upon the nature of the suspended matter, as pointed out above. The concentrations of dissolved and suspended matter can be controlled by testing gauge-glass samples as described in the Analytical and Testing Section.

Finally, priming may be prevented by the use of anti-foam agents. One of the oldest of these is castor oil, but the mechanism by which it prevents foaming is still uncertain. It is said to be due to a change in surface tension whereby the stability of the bubble films is impaired. It is to be noted that once priming has commenced it is useless to attempt to overcome it by the introduction into the boiler of successively small quantities of castor oil, since these are immediately carried away by the priming water. Hall[12] has used a number of anti-foaming materials, such as gallic and tannic acids and their compounds, oils such as sperm oil, castor oil, cotton-seed oil, and corn oil, waxes such as beeswax and carnauba wax, pyrogallol, tartaric and citric acids and their compounds, cholesterol and monohydric alcohols. He also found water insoluble anti-foaming material, of the general nature of the unsaturated hydrocarbons characteristic of asphalts, very effective. Results with some of the latter materials may be described more fully. From 6 to 8 lb. of active material were used per 1,000,000 lb. of raw water. Two 500-H.P. boilers of 150 lb. operating pressure were used, anti-foam being added only to one. Despite higher concentrations of hydroxide, sulphate, and suspended solids in the boiler water treated with anti-foam, the amount

PLATE 21. SOFTENING PLANT WITH GROUND CONTROL GEAR INSTALLED AT A WOOL DYEING WORKS

PLATE 22. BOBY SIMPLEX HEATER SOFTENER WITH PRESSURE SAND FILTERS. CAPACITY 168,000 GALLONS PER DAY

of water carried to purifier and thence discharged by trap was vastly less. To quote Hall: "The results during the addition of anti-foam were satisfactory. Trap discharges were not excessive in number or frequency, and superheat temperatures were steady. Immediately after stoppage of the anti-foam, however, the superheat temperature fluctuated violently and the boiler had to be blown down heavily to stop the foaming. Experiments were made on other operating boilers with indifferent or no success. A large enough shot of anti-foam seems capable of quieting boiling conditions temporarily, but, like the drug habit, leaves the user in a sad plight if successive shots are too infrequent or if continuous feed is insufficient or interrupted." It is evident, therefore, that the use of such anti-foaming material is of doubtful value. The most efficacious anti-foaming agent was castor oil, the effect being out of all proportion to the amount of castor oil added. Thus, Joseph and Hancock state that an emulsion of the oil (1 in 6 emulsion to 1 per cent soap solution) added to a boiler water in the proportion of 40 p.p.m. cured priming in concentrated waters and prevented its recurrence until a concentration of 9,800 p.p.m. was reached. A further 40 p.p.m. checked it at this stage also.

Hancock[18] has expressed the opinion that the effect of castor oil in suppressing priming is associated with its resistance to saponification under boiler water conditions. At the same time he presents further data on the use of castor oil as an anti-foam. A trial on a main line passenger engine with an anti-foam composition containing about 20 per cent castor oil indicated that correct and regular dosing is important, so that neither too much nor too little castor oil is used. A dosage of 0·4 lb. of the above composition per 1,000 gal. (=8 p.p.m. castor oil) allowed the concentration of salts in the boiler to be increased to 370 grains per gal., whereas the normal limit was 200 grains per gal. On the other hand, a dosage of 1 lb. per 1,000 gal. caused the boiler to prime at 110 grains per gal. salinity. On another occasion when 1 lb. of castor oil was pumped into the boiler (dose=85 p.p.m. castor oil) vigorous priming occurred after ¼ hour's running. The author has found castor oil effective in stopping priming on Lancashire boiler plant and on multiple effect evaporators.[19] In the case of the evaporator as little as 3 p.p.m. was often effective in stopping priming, even with a saline concentration of 15,000 grains per gal. More than 10 p.p.m. was never required. On the Lancashire boilers, 15 p.p.m. of castor oil was the amount used to prevent priming at saline concentrations up to 3,000 grains per gal. In each case the castor oil was added as such, and not as an emulsion or part of a composition. Castor oil and sunflower oil are also recommended as anti-foam agents by Gleim and Lissaievitch,[21] but they, too, report that there is an optimum quantity of oil above which the anti-foam effect is lessened, and which must be determined in each case. It is evident, therefore, that while castor oil can be an effective anti-foam, careful control of the treatment is necessary. With anti-foam treatments in general, it is often found that the anti-foam agent becomes ineffective after a while.

It may be of interest to note at this point that experimental work by Hancock[20] led him to classify oils and fats according to their behaviour in boiler priming. As a result of many experiments in alkaline and neutral waters he divided the more common oils and fats into three classes: (*a*) those with anti-priming properties, (*b*) those promoting priming, and (*c*) those with no effect either way. In the first class he placed castor, cotton-seed, and coconut oils and carnauba and paraffin waxes. Among the priming materials he placed linseed, olive, rape, and sesame oils and tallow, beeswax, and Japanese wax. In the "neutral" class he placed mineral oil and glycerine. Alkalinity produces fatty acids from the oils, and these acids were shown experimentally to be powerful promoters of priming. In agreement with this observation, it was observed that while coconut and cotton-seed oils were good anti-primers in neutral solution, in alkaline solution this property fell off.

Recently patented anti-foaming agents include fatty acid amides, such as stearamide,[13] and also water-insoluble metallic soaps dissolved in pine oil,[14] such as aluminium, barium, calcium, and zinc stearates and oleates. In the latter case, 1 per cent by weight, or less, is said to be sufficient when using 2–10 per cent by weight of the soap in making the solution.

An electrical method which has recently been patented consists in the insertion in the boiler water of a pair of copper electrodes. An electro-motive force imposed between the electrodes causes the production of copper and sulphate ions which neutralize the electric charge on the foam bubbles and are so claimed to destroy the foam.

Blow-down Control

As is well known, in order to keep the boiler water concentration within safe limits it is customary to blow down the boiler periodically, or in the case of a more modern plant there may be installed a system of continuous blow-down. In many cases blowing down a boiler is a rule-of-thumb procedure so far as the percentage of blow-down is concerned, blow-down cocks being opened periodically in a haphazard manner with no idea of the concentration of the water. However, the superiority of the more economic and scientific methods of controlling the boiler water concentration is rapidly being recognized, as is evidenced by the variety of continuous blow-down systems at the command of the modern engineer.

To emphasize the desirability of blowing-down it may be of interest to consider the effect of chemical treatment on the saline concentration of feed water. The two chief systems of feed water softening in present-day use are the lime-soda and zeolite or base-exchange treatments. Such chemical treatment may cause little or no reduction in the saline concentration of the treated water, as shown by the following table for a raw water supply treated by each of these systems:

TABLE 13

Raw Water		Water Resulting from Lime-soda treatment		Sodium Zeolite treatment	
Calcium carbonate ..	8·0	——		Sodium bicarbonate ..	8·5
Calcium sulphate ..	11·0	Sodium sulphate ..	11·5	Sodium sulphate ..	11·5
Magnesium carbonate	3·5	——		Sodium bicarbonate ..	4·4
Magnesium sulphate ..	8·0	Sodium sulphate ..	9·5	Sodium sulphate ..	9·5
Sodium chloride ..	10·0	Sodium chloride ..	10·0	Sodium chloride ..	10·0
Total solids	40·5		31·0		43·9

All figures in grains per gallon

In this example the lime-soda treatment has reduced the concentration of total solids by almost 25 per cent, but it will be observed that this is due to the ratio of temporary and permanent hardness and that the greater the permanent hardness the less will be this reduction in saline concentration. Similarly, the more magnesium salts there are present in the raw water the greater will be the increase in dissolved solid content caused by zeolite treatment. Finally, no chemical treatment yet devised has any influence at all on the sodium salts originally contained in the water. With a chemically pre-treated water, therefore, foaming and priming are liable to occur in the boiler unless the concentration is carefully controlled and the seriousness of solid carry-over during priming has already been illustrated by an example. Condensate is of great help in reducing the rate of concentration, and the quantity of water required to be blown from a boiler is reduced in direct proportion to the quantity of condensate available. But if the make-up is evaporated, care must be exercised even in this case, since the evaporator may prime or a small condenser leakage obtain. Even the small quantity of solids which thus passes to the boiler will rapidly increase to undesirable proportions. This can be illustrated by considering the case of a boiler of 50,000 lb. capacity at normal working level and with an average evaporation of 40,000 lb. per hour. Suppose the make-up is 5 per cent and that it contains 10 grains per gal. of soluble salts. For this illustration assume that the initial concentration of the boiler water is nil and determine how long is required for the boiler water concentration to reach 100 grains per gal. The volume of make-up per hour is 200 gal., so that 2,000 grains of salts enter the boiler each hour. Hence the number of hours steaming required to reach a concentration of 100 grains per gal. is 250 hours. That is about 1½ weeks' continuous steaming.

Foaming and priming have been mentioned as two unwelcome consequences of uncontrolled concentration of dissolved and suspended matter in the boiler. Another point is that salts in solution increase the density of the water and thus militate against free circulation. Such an effect may not be appreciable in most cases, but that it may on occasion be severe is borne out by a case described some years ago. In this case the boiler was laid off for internal examination and was found to contain almost a solid mass of crystals. The feed had been treated for acidity without care being taken to keep down the density of the water in the boiler. The result was that circulation was practically stopped and overheating of the plates occurred.

Deconcentration of the boiler water can be effected either by instituting a regular periodical blow-down of predetermined amount or by adopting a continuous blow-down system, the quantity of water run from the boiler being regulated by a suitable valve. The water which is blown down is at the temperature and pressure of the boiler, and consequently in the former operation both heat and pressure are lost, the process being most uneconomical. For example, it can be shown that with a boiler working at 200 lb. per sq. in. pressure and using coal of 12,000 B.Th.U. calorific value, every 1,000 gal. of water blown down contains heat equivalent to about 270 lb. of fuel. Any water unnecessarily blown down therefore means a considerable wastage of coal. This brings out the advantage of continuous blow-down systems. In these the operation of blow-down continues all the time the boiler is working, but the amount of water removed in any one period is small. The heat content of the blow-down is not lost, as it is transferred almost completely to the feed water. These systems will be discussed in greater detail later. In a large number of cases the amount of water that requires to be blown down to maintain a safe concentration of total solids is small, and does not justify the expense of a continuous blow-down appliance. In such a case a periodical blow-down should be instituted and the following would be the most economical way of carrying out this operation.

Intermittent Blow-Down

In the first place, the limiting concentrations of dissolved and suspended material should be decided upon. Now while a figure of 1,000 grains gallon is often taken for Lancashire boilers, and half this amount for water-tube installations, it really cannot be arbitrarily fixed in this way owing to the varying conditions encountered. The figures will depend on various factors such as boiler design, operating pressure, and boiler rating, so that experience with the particular boiler in question is probably the best guide in this respect. The percentage of blow-down necessary to avoid exceeding this limit can then be calculated. The frequency of blow-down naturally enters into the calculation, and to a certain extent this can be arranged as is most convenient.

For the purpose of deducing formulae the several variables can be denoted as follows:

E = average rate of evaporation of the boiler in gallons per hour.

F = concentration of dissolved solids in the feed water in grains per gallon.

C = maximum concentration of total solids to be attained in the boiler, in grains per gallon.

B = the quantity of water in gallons per hour to be blown from the boiler.

To keep the concentration of the boiler water at a certain limit, the quantity of dissolved solids blown from the boiler during a given period must equal the quantity entering the boiler in the same time. The quantity of dissolved solids entering a boiler in, say, an hour, is given by the product EF. The quantity leaving the boiler during blow-down is BC. Therefore to maintain the concentration at the value C, we have $BC = EF$, or $B = \frac{EF}{C}$. The concentration C will have been predetermined, E and F are known, so that B is readily found. This gives the extent of blow-down per hour, which is what is required for continuous blow-downs, but for periodical blow-downs information is required regarding the frequency of the operation. If N is taken as the number of hours steaming at the rate of evaporation between successive blow-downs, then it follows that $NEF = BC$, or $B = \frac{NEF}{C}$. Obviously, having determined the value of B for a given case, it is unnecessary to commence the periodical blow-down until the limiting concentration agreed upon has actually been reached. The time to commence can readily be found in a similar manner to that adopted above. Thus, if n is the number of hours steaming necessary for the boiler water to increase in concentration from zero (i.e., when the boiler is first put back on the line) to C, and W is the capacity of the boiler in gallons at normal working level, then $C = \frac{nEF}{W}$, whence $n = \frac{CW}{EF}$. Of course, the time and blow-down given by these formulae should be checked periodically by determining the dissolved and suspended matter in the boiler water. Such checks are advisable, since the above formulae are based on the assumption that the feed water supply and the evaporation rate remain constant at the values used. Any variations cause differences in the calculated values for the total solids and those actually found. Such variations will be the more liable to cancel out the longer the period over which the formulae is applied. To illustrate, the formulae $C = \frac{nEF}{W}$ gives the concentration of the boiler after n hours steaming. The greater the value of n the more nearly will C, as calculated, coincide with the actual value as determined.

The fuel loss due to blow-down, when no heat is recovered by heaters, may be calculated as follows: $W = \frac{10B(T\text{-}60)}{C}$ where B is the boiler blow-down in gallons per hour, T is the temperature of the water at the pressure of the boiler, C is the calorific value of the coal used in B.Th.U. per lb., and W is the equivalent weight of coal in lbs. A considerable part of this heat can be recovered by passing the blow-down water through a heat exchanger or suitable coil arrangement to reduce its temperature to about 80°–100° F. before it passes to the drain.

Continuous Blow-Down Systems

In the case of boilers operating at high evaporation rates, even if the percentage of make-up is kept low, the amount of blow-down necessary may be great. In comparison with the small water capacity of drums, this amount of blow-down is usually such that it cannot be blown down at one time without a dangerous reduction in water level, and the alternative intermittent blow-down would involve very frequent blow-down, which militates seriously against the efficient operation of the plant. The continuous process was therefore developed in which the operation of blowing-down continues all the time the boiler is working, a relatively small quantity of water being removed per unit of time at a constant rate. The advantages of such a system over periodical or intermittent blow-down are: (*a*) The heat in the blow-down may be utilized either for pre-heating purposes or for the operation of hot-process softeners, whichever may be the most suitable; (*b*) the constant removal of dissolved and suspended material from the boiler permits the maintenance of uniform concentration and reduces scale formation and sludge accumulation.

The water removed from the boiler contains, in addition to the sludge, sodium sulphate, etc., which it is desirable to remove, other substances, such as caustic soda and sodium carbonate, which, together with the sensible heat of the blow-down, are of considerable value. Kestner was one of the first who studied the problem of utilizing the heat and soda alkalis in the boiler blow-down. In his system, as originally designed, the blow-down water was merely freed from suspended material and then returned to the boilers together with the softened make-up. The make-up was chemically softened in a hot-process softener, the blow-down supplying the heat, and, of course, the alkalis in the blow-down served to treat the make-up. However, this arrangement did not reduce the concentration of the permanently soluble salts in the boiler water, which could only result if the blow-down were passed through a heat interchanger and then discharged to waste.

A similar system is the Neckar process. In this case there is additional equipment to prevent the increase of concentration of the soluble salts above any prearranged limit. A proportion of the blow-down is by-passed into a closed vessel which is open to the atmosphere through a vented drain, and the heat liberated from the boiler water, due to drop in pressure, produces steam. This is returned to the boilers as pure condensate. The residue of the boiler water then passes through tubes in the lower part of the vessel on its way to the drain, and gives up the remainder of its heat to the water on the other side of the tubes. For the complete equipment it is claimed that only about 5 per cent of the total heat in the blow-down is lost, and that from 15–20 per cent of the blow-down is returned to the boiler as condensate. The actual percentage of blow-down recovered as condensate depends upon the boiler pressure and may be easily calculated as follows: If T is the temperature of the boiler water, then the heat released in the expansion chamber is T—212, whence the condensed portion of the water is $\frac{T-212}{966} \times 100$ per cent. At 200 lb. boiler pressure this would amount to nearly 18 per cent. The only chemical normally used in the process is soda ash, which is added in quantities sufficient to react with the permanent hardness salts in the untreated water. The bicarbonates in the untreated water are removed by the heat and by the caustic soda in the boiler water enter-

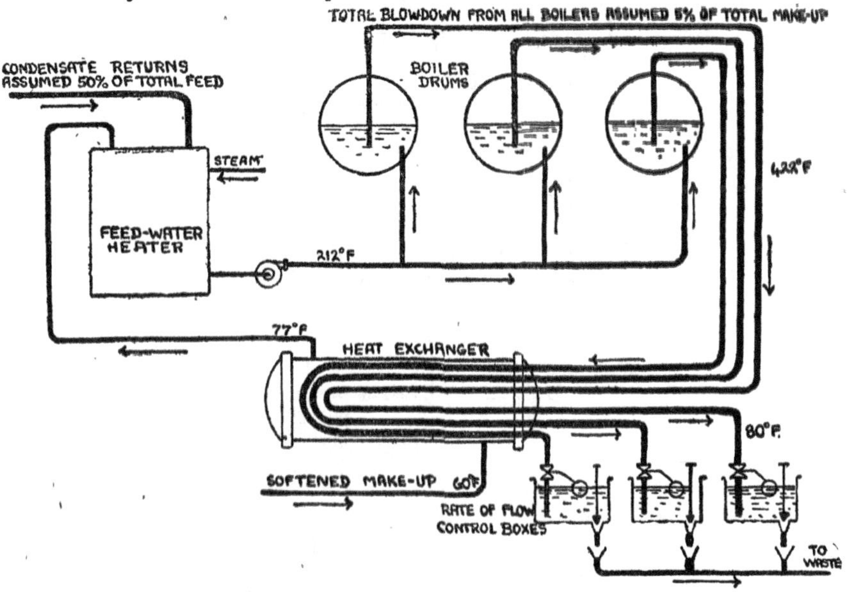

FIG. 30. Continuous blow-down system with heat recovery in exchanger (Permutit)

ing the softener. In the Boby continuous blow-down system also a portion of the blow-down is allowed to flash off to steam, the condensate being returned to the boilers with the feed, and the remainder of the blow-down passing through a heat interchanger before being run to the drain. In the Permutit system of continuous blow-down no attempt is made to return the blow-down water to the boilers; the available heat is extracted and the blow-down discharged to waste.

The selection of one of these different systems of continuous blow-down for given conditions depends on various factors such as type of control preferred, necessity for condensate recovery, boiler operating pressure, etc. As plant pressures increase arrangements naturally become more elaborate, but for low and moderate pressures quite straightforward systems can be used. This latter is exemplified by the "Permutit" system illustrated in Fig. 30. The circuit here is quite straightforward, in that the blow-down simply passes through a heat exchanger of the multi-stream type for the extraction of heat before discharge to drain. The heat extracted is used to raise the temperature of the feed water on its way to the feed heater, and the rate of flow of the blow-down from each boiler is individually controlled as required and very simply. If a single-stream heat exchanger is used in place of the multi-stream heat exchanger in Fig. 30, then it will be necessary to combine the several blow-down streams before they enter the exchanger. In these circumstances control is still exercised after the exchanger, since control prior to it will tend to cause flashing

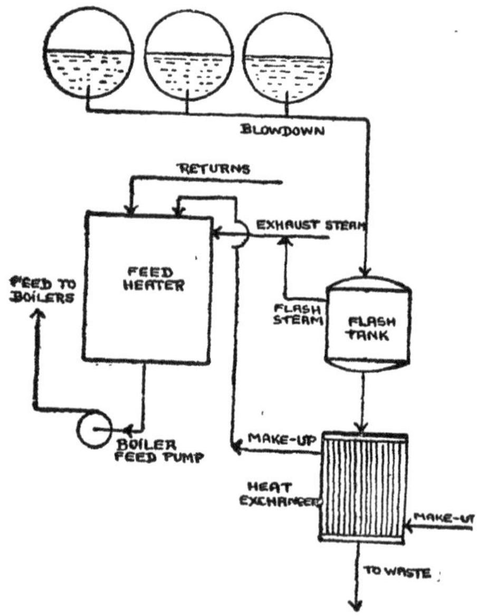

FIG. 31. Continuous blow-down system with heat recovery in flash tank and heat exchanger.

and danger of water-hammer. The flow from the several boilers can, however, be balanced by valves before the exchanger providing the over-riding control comes later.

The "flash" system layout is very similar as shown in Fig. 31. The blow-down is first passed into the flash tank located between the boilers and heat exchanger, and the flash steam utilized in any of various ways, after which the blow-down is further cooled in the heat exchanger by means of the make-up on its way to the feed heater.

Both the foregoing layouts are quite simple and have been adopted very successfully by the ordinary small industrial plant in many cases. Numerous variations are possible to meet individual conditions. It will be noted that all the heat is recovered for use in the open feed heater, either by direct exchange with the make-up or in part by admission of flash steam as well to supplement the exhaust steam fed to the heater. The flash steam can be recovered as condensate, which is fed to the boilers or softening plant, or it can be passed to a low-pressure steam main. The heat exchanger can be located so that the heat in the blow-down passes to the raw water on its way to the softening plant, or part of the blow-down can be fed to the softening plant for utilization of the alkalies for softening purposes (one battery of boilers operating at 160 lb./sq. in. cut chemical costs over 20 per cent this way). Variations on these lines are needed to enable the blow-down recovery to fit into the heat balance of the plant; for instance, they may be necessary to effect heat savings on a plant where the feed heater is adequately supplied with heat from the available exhaust steam.

An alternative layout is that shown in Fig. 32, where two heat exchangers are used. The high-duty exchanger is installed between the

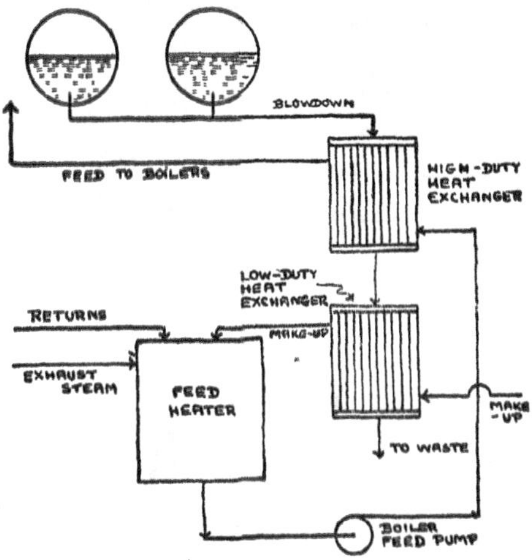

FIG. 32. Continuous blow-down system with heat recovery in high- and low-duty heat exchangers.

feed pump and the boilers, so that the blow-down passes through this exchanger first, and is cooled down to near the final feed-water temperature. The partially cooled blow-down then passes on to the low-duty heat exchanger, where the remaining heat is given up to the make-up on the way to the feed heater. This type of layout is useful where all the heat in the blow-down cannot be transferred to the make-up. Part of the heat is absorbed by the final feed water, but a part is left to supplement the exhaust steam to the feed heater; any slight deficiency in the exhaust steam available can thus be made up.

This type of system can be elaborated to include a flash tank on the lines of Fig. 31, using the flash stream to produce condensate or feed into a low-pressure steam main, with the flash tank between the two exchangers or before the high-duty exchanger, as operating conditions and the plant heat balance dictate. Conditions on a plant operating at 650 lb. per sq. in. with this type of hook-up are shown in Fig. 33. The blow-down water at 515° F. first flows through a high-duty heat exchanger, where its temperature is reduced to 360° F.; heat saving is effected because the feed water entering this exchanger is already at a temperature higher than that available from any other source of heat on this plant. After the high-duty exchanger the blow-down passes into

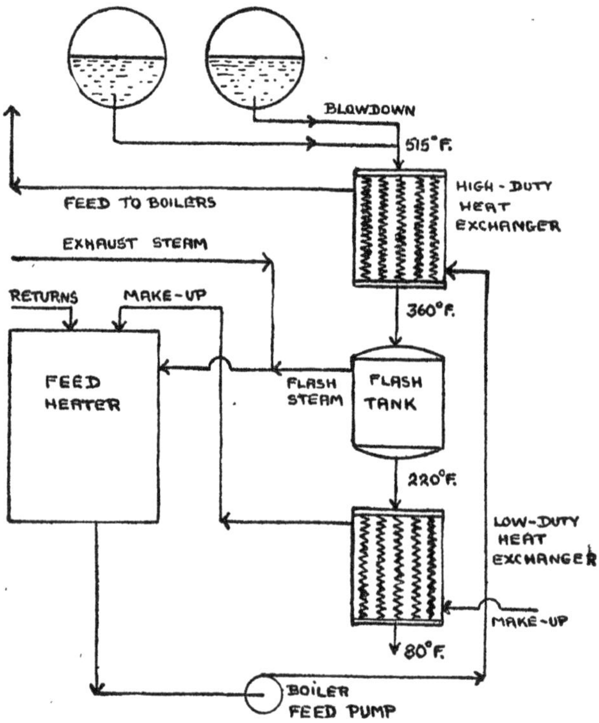

FIG. 33. Continuous blow-down arrangements on plant operating at 650 lb. pressure.

a flash tank, where the pressure is reduced to 7 lb. gauge; 15 per cent by weight of flash steam is thus produced, and this is admitted to a de-aerating feed heater together with the exhaust steam from fan turbines. At the exit from the flash tank the blow-down is at a temperature of approximately 220° F., and this is further reduced to about 80° F. (which is only a few degrees above the make-up supply to the plant) in the low-duty exchanger.

At higher operating pressures it may be necessary to use two flash tanks in series, as shown in Fig. 34. In such an arrangement flash steam is first taken off the blow-down in a flash tank coupled to a low-pressure

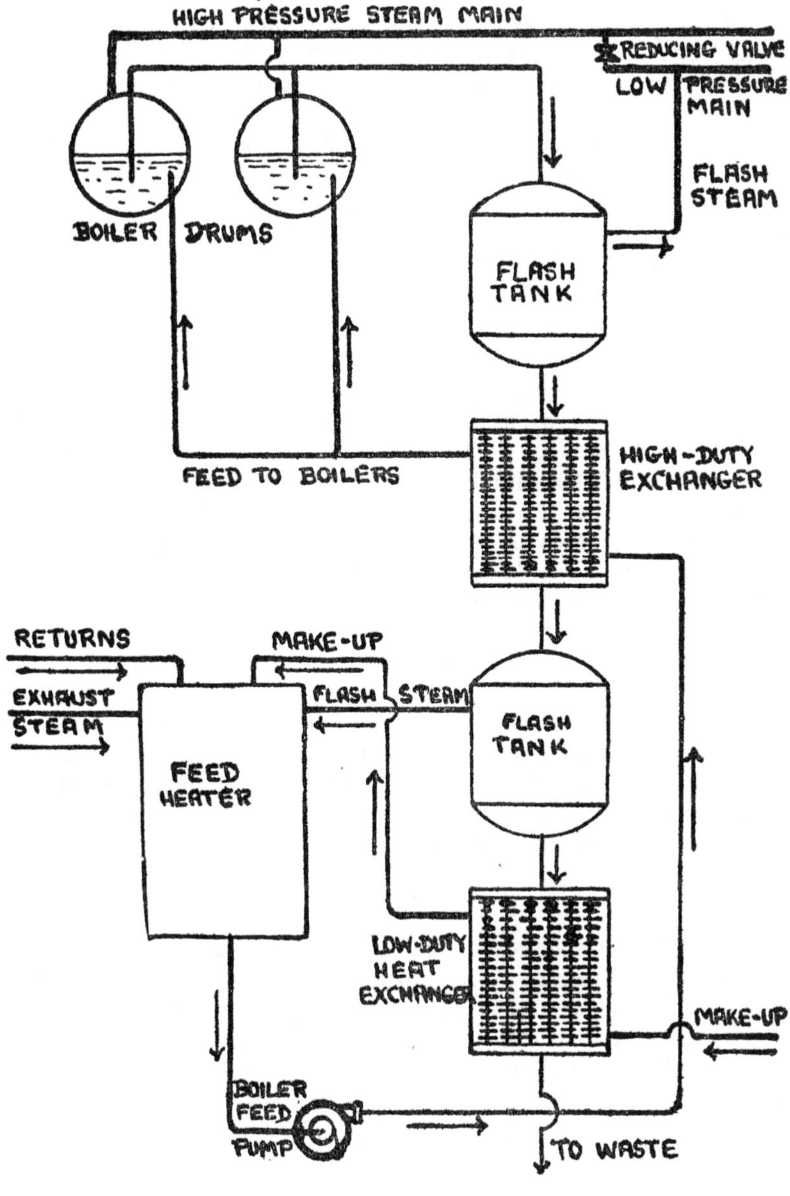

FIG. 34. Continuous blow-down system for plants operating two steam pressures.

steam main, and the partly cooled blow-down is then available for heating feed water in a high-duty exchanger. Further heat recovery is effected by passing the blow-down through a second flash tank and a low-duty heat exchanger. The writer recalls a battery of 1,640 lb./sq. in. water-tube boilers on the Continent which operate a system something on these lines.

It will be evident that the type of hook-up sketched in Fig. 34 can also be adopted on smaller plants where two different pressures are operated, provided the heat balance of the plant permits. For example, in a plant operating at 350 lb./sq. in. and 80 lb./sq. in. pressures, the flash tank would be set at 80 lb./sq. in. by connecting the steam outlet to the 80 lb. main. The non-flashed blow-down liquor from the flash tank can next pass through the high-duty exchanger, following up through the second flash tank and low-duty exchanger. This type of lay-out effects heat savings (*a*) partly by addition of heat to the final feed water, (*b*) partly by addition of heat to the make-up water, and (*c*) partly by provision of flash steam to the low-pressure steam main. The flexibility thus permitted in heat recovery enables direct savings to be achieved under various heat balance conditions. For example, the hook-up overcomes difficulties which arise if ample supplies of exhaust steam are available for feed heating.

Heat recovery arrangements modelled on the foregoing basic layouts are adopted with necessary individual modifications on all modern steam-raising plants, particularly where operating pressures are high. It is due to meticulous attention to such details that modern power stations and high-pressure plants operate with their high standard of fuel efficiency. The ordinary industrial boiler plant will profit materially by following their example as far as possible. In the field of blow-down heat recovery very considerable improvements could certainly be effected on most works boiler plants.

Use of Condensate and Blending of Waters

The amount of condensate available is of great assistance in reducing the rate of concentration, since it will be evident that the quantity of water which has to be blown from the boiler to keep a certain concentration is reduced in direct proportion to the quantity of condensate which is used. Modern condensing plant with its high efficiency may therefore enable a source of feed supply to be used which without the addition of condensate would be totally unfit for boiler feed purposes owing to the high dissolved solid content. A similar expedient is the blending of two or more natural waters. One water high in dissolved solids may sometimes be mixed with another low in dissolved solids to give a blended water, completely softened, and containing only a moderate amount of dissolved solids. This renders usable the supply with the high solid content and also economizes from the aspect of chemicals required for softening purposes. This expedient depends upon the existence of two convenient suitable sources of water, but can often be adopted in the case of pit waters, where two or more different waters are sometimes obtained from different seams of the same colliery. The operation of blending and its advantages are dealt with more fully in the section on scale formation.

Removal of Oil from Water

When condensed steam alone or mixed with natural water is used for boiler feeding it usually contains oily and greasy matter, and this must be removed. It has been shown by British Navy engineers that the presence of even minute traces of oil may have serious consequences, and most insurance companies now refuse to insure boilers which are fed with oily feed water.

Oil may be present in feed water in two forms—in suspension and in emulsion. Emulsified oil is considered to be one of the contributory causes of priming, while suspended oil leads to overheating. The influence of oil in foaming and priming troubles has been discussed. Overheating troubles are caused by the tendency of oil to stick to the plates and, in fact, bulging of the plates or tubes may occur through the joint action of internal pressure and plasticity of the metal caused by local overheating. Furthermore, the oil chars under the action of heat, leaving a carbonaceous mass behind on the plate or tube. This readily collects more oil, so that further overheating takes place. The very minute quantity of oil that is sufficient to cause this is not always appreciated. A layer of oil of not more than ½ mm. thickness in contact with the heated boiler wall is equivalent in heat insulating power to a 5 mm. thickness of boiler scale.

Of course, the source of oil in condensed steam is the oil used for lubricating the cylinder. The viscosity of oil is so reduced by the heat that it is readily blown out by the exhaust in the form of a fine spray mixed with the steam. In the case of saturated steam the oil is carried as thin films of oil surrounding the drops of water. Oil suspended in this way may be separated mechanically. The steam may be passed through a baffle-plate separator, the water and oil being collected and run off together. This serves very well for oil which is present as a spray. There is still that portion which is vaporized and which will be condensed with the steam.

Modern lubricating oils are mixtures of hydrocarbons of various boiling points, the lowest boiling constituents of which are readily vaporized at the temperature of steam. We cannot expect, therefore, that separators placed in the steam pipe will remove all the oil, however good they may be. Some tests made at the Massachusetts Institute of Technology on baffle-plate oil separators may be quoted in this connection. These showed[15] that the amount of oil separated in this way is about 75 per cent of the total oil contained in the steam. The remaining 25 per cent of oil goes to the condenser and is precipitated in the water in the form of very minute globules. However, there are plants on the market which separate mechanically and more or less completely the suspended oil from the condensed water. Even here, though, there is still the problem of removing the emulsified oil. It is unfortunately not generally appreciated that emulsified oil cannot be removed by filtration. The drops of oil are extremely small, averaging only 0·00005 of an inch in diameter, and it is on account of this extremely small diameter that the difficulty of separation arises. Such an emulsion is extraordinarily stable, and samples have been kept for periods of years without showing any signs whatever of separating into their constituents. The best way of removing such oil is by coagulation and subsequent filtration. Such emulsified oil cannot be removed by simple filtration through filters ; even though the filter may be effective for removal of suspensions of the finest mud.

Methods of De-Oiling Water

The methods of removing oil from water may be classified as mechanical, chemical, and electrical. In mechanical methods the oil and water are usually separated by means of baffle-plates, the water being driven in a zig-zag direction through depositing tanks, the separation of oil at the upper surface usually being hastened by aeration or pressure. These plants, as already mentioned, will not deal with emulsified oil. One outfit consists of a tank having at one end a float chamber and at the other end a discharge pipe. The tank contains a number of vertical baffles by means of which the water and oil entering at the bottom are caused to separate, the oil rising until it fills the float chamber. This contains a float so adjusted that it opens a discharge valve when the chamber is full. When the oil has left the chamber it fills again, and the process continues automatically until there is no more water-oil mixture flowing into the bottom of the tank. In another separator the oily steam is first passed into a concentrant, in which the finest globules of oil are coalesced. The steam with these larger globules of oil then passes over a series of steel baffle-plates which cause the actual separation of the oil and steam.

The Davis-Perrett oil separator operates by electrolysis of the condensate. The hot condensate is passed through a tank divided into sections which are subdivided by iron plates. These plates serve as electrodes and about 50 volts potential is used across each plate. The feed is mechanically filtered after coagulation has occurred, wood wool or sand filters being used. The plates last about two or three years and may be cleaned of deposit (even when badly covered) by reversing the current. A Dutch engineer has described a similar method of dealing with emulsified oil. The tiny globules of oil are made to coalesce by passing an electric current through the water. He uses 110 volts continuous current while the water flows through the feed pipe The electrically treated water is then filtered by means of a pressure filter. This consists of perforated cylinders covered with cloth and enclosed in a suitable vessel. Since the globules of oil are increased in size by the electric current, this filter removes all the oil. A pressure gauge on the filter indicates when the cloths need changing.

An emulsion is said to be stable if it does not separate readily into its constituents, the degree of stability being measured by the time it takes to separate. Separation is caused by the coalescing of the minute globules forming the emulsion. If these droplets are charged with electricity of the same sign they will repel one another and not coalesce. The emulsion will then be stable. In many cases where the stability of an emulsion is due to the electrical charges upon the droplets, the emulsion may be broken by the addition of suitable electrolytes. The added ions neutralize the charges upon the oil particles and so lead to the breakdown of the emulsion. Similarly, small particles might be precipitated upon the surface of the oil globules if they were charged with electricity of a sign opposite to that on the oil droplet. The electrical charges would cause an attractive force. If, then, oil globules will collect small particles of certain solids around them, the filtering of an oil emulsion through a large mass of these substances should filter off the oil and give a clear filtrate. This is exactly what was found to be the case. Filter media formed of the following substances gave a clear filtrate—hydroxides of iron, aluminium, chromium and calcium, barium carbonate, bone black, spongy iron and finely ground coke. Hatschek[16] also found that filtration through magnesium or calcium carbonate would also remove all the oil. Ground coke has often been used for filter beds, but attempts should not be made to give it some regenerating treatment so that it can be used as a filter again. It is advisable to use the old filter coke and prepare a fresh filter bed. For normal purposes a bed of 2–4 tons of coke should be sufficient, the material being changed every month or six weeks. Naturally, filtration should be downwards to counter the lighter nature of the oil. Some arrangement should be provided to prevent dust being carried off with the water, and coke should never be used in an acid condition. An alkaline wash is required until a coke which is acid tests properly neutral. The best coke is a fine, porous product, about 1 in. grade.

More convenient for practical purposes is the method of removal by the addition to the water of aluminoferric and soda ash or caustic soda, followed by filtration. It is important that the reagents should be thoroughly mixed with the water. Aluminoferric is an impure form of aluminium sulphate and reacts with caustic soda and sodium carbonate in accordance with the following equations:

$$\underset{\text{Caustic soda}}{6NaOH} + \underset{\text{Aluminium sulphate}}{Al_2(SO_4)_3} = \underset{\text{Aluminium hydroxide}}{Al_2(OH)_6} + \underset{\text{Sodium sulphate}}{3\,Na_2SO_4}$$

$$\underset{\text{Sodium carbonate}}{3Na_2CO_3} + \underset{\text{Aluminium sulphate}}{Al_2(SO_4)_3} + \underset{\text{Water}}{3H_2O} = \underset{\text{Aluminium hydroxide}}{Al_2(OH)_6} + \underset{\text{Sodium sulphate}}{3Na_2SO_4} + \underset{\text{Carbon dioxide}}{3CO_2}$$

The oily globules become entangled in the gelatinous precipitate of aluminium hydroxide, which is removed by sedimentation and filtration. Complete removal of the oil is thus effected. It will be noted that the coagulant here is aluminium hydroxide, which, as already mentioned, is capable of neutralizing the electric charges on the oil droplets. The effectiveness of this method is therefore probably due in part to such action. The oil globules and the aluminium hydroxide are attracted to one another by their electrical charges and the bulky nature of the aluminium hydroxide facilitates sedimentation and subsequent filtration. On account of its corrosive properties, it is important to avoid the presence of any undecomposed aluminium sulphate in the treated water, which should contain a slight excess of soda or caustic alkalinity.

In the Harris Anderson de-oiling plant, two fractional amounts of the total water supply are separated automatically and made into solutions of caustic soda and aluminium sulphate respectively. These then mix with the water forming a flocculent precipitate of aluminium hydroxide, which with its entrained oil is removed by a series of filters of wood wool. The two fractions so converted into reagents bear a constant relationship to the supply, being obtained by a turbine worked by the incoming water.

A recent development consists in the use of sodium aluminate. The arrangement is to have two solution tanks, one for filter alum or aluminium sulphate and the other for sodium aluminate, each feeding small constant-level tanks fitted with taps. The quantities of chemicals are adjusted to leave a slight alkalinity in the effluent. Filter alum or aluminoferric is acid, while sodium aluminate is alkaline; therefore these two together quickly neutralize each other to form sodium sulphate and liberate gelatinous aluminium hydroxide. This results in a double coagulation which is very effective in removing the oil.

It should be emphasized that sodium aluminate contains alumina and is also alkaline, so that it is more useful than caustic soda or soda ash in the oil-removal plant. By using a slight excess of sodium aluminate in the condensate, therefore, the corrosive effect of filter alum is eliminated.

Oil removal in this way is exceptionally efficient, and it has been stated that the method can be adapted to remove quantities as low even as 0·1 grain of oil per gal.

Another material which has recently been applied to de-oiling work is activated carbon. Experiments with this material have been described in which the condensate from a reciprocating steam engine, with an oil content of 5–6 grams per cubic metre, was treated and reduced to 0·8 gram per cubic metre, i.e., to about $\frac{1}{2}$ grain per gal. At this point the condensate was perfectly clear and colourless. By performing the operation of filtering the condensate through the activated carbon in the hot the residual oil in the water can be brought down to less than 0·1 grain per gal.[17] According to Kagan,[23] active carbon is more effi-

Plate 23. WATER SOFTENING AND DE-OILING PLANT AT A PAPER MILL

PLATE 24. SOFTENING PLANT AT FORD MOTOR WORKS, DAGENHAM

cient than aluminium hydroxide for removal of oil from condensed steam. Birch carbon is said to be more suitable than peat carbon, and he also recommends that the condensate be purified mechanically before filtering through active carbon.

REFERENCES

(1) Joseph & Hancock, " Chem. & Ind.," 1927, p. 315T.
(2) Lord Rayleigh, " Chem. News," 1890, p. 52.
(3) Tatlock & Thompson, " Chem. & Ind.," 1904, p. 428.
(4) Koyl, " Railroad Gazette," 1900, p. 663, and 1902, p. 423.
(5) McGill, " Chem. & Ind.," 1904, p. 351.
(6) Foulk, " Ind. & Eng. Chem.," 1924, p. 1121.
(7) Behrman, ibid., 1929, p. 817.
(8) Hancock, " Chem. & Ind.," 1930, p. 369.
(9) Bancroft, " Applied Colloid Chemistry," McGraw Hill Book Co., 1921, p. 268.
(10) Powell, " Power," 1926, p. 330.
(11) Beale & Stevens, " Chem. & Ind.," 1931, p. 307T.
(12) Hall, Can. Patent, 284, 637; see also Hall, " Ind. & Eng. Chem.," 1929, p. 824.
(13) U.S. Patent, 1,892, 857 (1933).
(14) British Patent, No. 383, 293 (1933).
(15) Ellis, " Chem. & Ind.," 1912, p. 909.
(16) Hatschek, ibid., 1910, p. 125T.
(17) Jaenicke, German Patent, No. 601, 421.
(18) " Engineering & Boiler Hse. Review," May 1938, p. 766.
(19) Matthews, " Eng. & Boiler Hse. Rev.," December 1938, 388.
(20) Hancock, " Journ. Soc. Chem. Ind.," 1930, *37*, 369T.
(21) Gleim & Lissaievitch, " Chim. et Industr.," 1938, *39*, 1097.
(22) Foulk, " Engineering Experiment Station News," Ohio State University, Oct. and Dec. 1937.
(23) Kagan, " Chim. et Industr.," 1938, *39*, 1097.

PART FIVE

ANALYSIS AND ROUTINE TESTING

THE PRINCIPLES UNDERLYING THE SOFTENING AND CONDITIONING OF boiler feed water to prevent scale formation, corrosion, and embrittlement have already been described. It will be appreciated from those discussions that if the maximum benefit is to be derived from softening and conditioning the processes must be carefully controlled, otherwise uneconomical operation of softening plant—and possibly dangerous feed conditions—will result.

The routine tests required to maintain this control are simple and can easily be carried out after a little practice has developed the necessary manipulative skill. Before proceeding to describe these tests, however, some general points should be emphasized.

In the first place cleanliness is essential. The test bench and all apparatus should be kept scrupulously clean and dry. All measures, bottles, dishes, flasks, and other glass apparatus should be thoroughly washed after use, preferably with hot water, allowed to drain, and then rinsed with the sample to be tested. Another point is lighting. The accuracy of analysis often depends upon observation of the change of colour of an indicator, so that tests should be carried out in a well-lighted place, and if possible in daylight.

Sampling

It is of little use to carry out analyses carefully and accurately if the samples are not representative. For instance, it is of little use to take a sample from a softener just after a change in the addition of reagents and expect a change in results. Time must be allowed for the change to become effective, and this will vary with individual plants. Experience should dictate how long is required before a sample can be taken, but generally 4–6 hours will be adequate.

The most suitable bottles for sampling purposes are glass-stoppered, clear-glass Winchester quarts. These should be kept scrupulously clean and when a sample is being taken the bottle should be rinsed out several times with the water to be sampled. If possible, samples should be collected directly in the bottles, but if this is not feasible, metal vessels should not be used. The bottles should be filled to within an inch or so of the stopper, this amount of air space being advisable to prevent possible bursting of the bottle. In collecting samples from rivers, tanks, etc., the sampling vessel should be plunged into the water so that its mouth is at least 2 in. below the surface of the water. This avoids surface contamination. Samples should be taken from taps only after the water has been running a sufficient time to remove all water which has been standing in the pipes. Samples for estimates of dissolved gases should not be agitated while being collected and should be kept as cool as possible until the analysis is made; early analysis is advisable. Boiled samples drawn from the water gauge cock should be collected after flushing out and care should be taken to close the top steam cock and so avoid dilution of the sample with condensed steam. The water may be collected in a clean bucket and allowed to stand in an earthenware jar until cool enough to transfer to a bottle. The ideal is to cool the sample as it is being drawn by passing it through a metal cooling coil. All samples should, of course, be clearly labelled with full particulars.

Standard Solutions

Particulars for making-up these solutions are given in the appendix. They may also be purchased ready made-up from the various reagent houses. In any case they should be tested periodically and their accuracy checked, say, every two or three months. This is particularly necessary (and in fact every month) with the soap solution, which contains alcohol liable to evaporate and so leave a stronger solution. For checking the soap solution a standard hard water is required; this may be of 14 Clark degrees hardness and is prepared as described in the appendix. It should be used exactly as in testing water samples. If solutions are too strong they should be diluted with distilled water to the required strength (alcohol used for the soap solution) or a correction factor may be calculated for correcting all titration readings.

Frequency of Analyses

Samples of untreated water should be analysed once a day unless the supply shows little signs of variation, in which case once a week will suffice. The frequency of analysis of the treated water will, of course, depend upon the suitability or otherwise of the operation of the plant.

Units

The results may be expressed in various units. The usual "degree" of hardness represents calcium and/or magnesium salts equivalent to 1 grain of calcium carbonate, which is the same as 1 part in 70,000. This is also sometimes called Clark's degree. Another way of expressing hardness or alkalinity which is becoming more general nowadays is the Continental degree, which is 1 part of calcium carbonate per 100,000. This, of course, is actually 0·7 grain per gal., or 0·7 of Clark's "degree." Yet another expression is parts per million. It should be realized that while hardness or alkalinity may be expressed as so much calcium carbonate it does not necessarily mean that the water contains calcium in solution as carbonate. Calcium carbonate is practically insoluble and the actual amounts of other calcium and magnesium salts which are in solution are expressed in terms of the equivalent amounts of calcium carbonate merely for convenience and to provide a common basis for comparison of analyses. Throughout this book the above expressions will be interpreted as set out here, and in so far as hardness and alkalinity are concerned the Continental degree, or parts per 100,000, will be adopted unless otherwise stated.

CONTROL TESTS FOR SOFTENING

The chief controllable factors in water softening are the hardness and alkalinity. Several methods are available for these tests.

(*a*) Total Hardness of Raw Water

The usual routine method of determining the hardness of water is by titration of a known volume of the water with a standard soap solution. It is well known that hard water uses up a certain amount of soap before it will produce a lather. The quantity of soap solution of a fixed and definite strength which is required to form a permanent lather on the water gives a measure of the hardness of the water.

The standard soap solution used may be either Clark's or Wanklyn's. Details for making up these solutions are given in the appendix, but it should be emphasized that the preparation and standardization is a skilled operation and if no chemist is employed, it will be advisable to obtain a stock from one of the reagent firms. It may be mentioned also that large stocks should not be carried as these solutions do not keep good for more than a few months. The stock of soap solution may be checked from time to time with a standard calcium chloride solution (as described in the appendix) to determine when fresh soap solution is required. In carrying out a hardness test as below, 50 ml. (cubic cms.) of water should be taken with Clark's soap solution, and 70 ml. of water if Wanklyn's soap solution is used.

This quantity of water is poured into a glass-stoppered bottle having a capacity of about 250 ml. The soap solution is then run in slowly say rather less than 1 ml. at a time—until a lather shows signs of forming on the surface. The bottle is, of course, stoppered and shaken between each addition. When signs of a lather are observed smaller quantities of soap solution are added at each time. This is continued until a lather is formed which remains for five minutes when the bottle is lying on its side. With waters containing magnesia a false lather may appear before the final end-point. Care should be taken not to confuse this with the true result. If there is doubt about a lather, take the burette reading and then add a further 1 ml. of soap solution (0.5 ml. at a time) and shake the bottle again. If the lather is the true one it will remain, but a false one will disappear. In the latter case the titration should be continued.

When Wanklyn's soap solution is used, a correction must be deducted from the burette reading to allow for the soap required to give a lather with pure water of zero hardness. This blank test can be made in the same way as above with 70 ml. of distilled water—usually rather more than 0·5 ml. is required to give a lather with distilled water. The remainder, i.e., the burette reading less the blank reading, represents degrees of hardness, or the amount of total hardness in grains per gal. This is parts per 70,000 and is easily converted to parts per 100,000 by dividing by 0·7. When Clark's soap solution is used the hardness corresponding to the amount of soap used is obtained from the table of Clark's degrees of hardness given in the appendix.

With both soap solutions, if the water is above 14 Clark degrees hardness, it should first be diluted with distilled water before testing. Thus, 50 ml. may be diluted to 100 ml. and then 70 ml. titrated with Wanklyn's soap solution. Of course, the final reading, less the blank, should be multiplied by two to give the total hardness.

The above soap method is very suitable for general control tests, but some operators may prefer the procedure devised by Blacher.[1] He used a standard potassium palmitate solution, made up and standardized as described in the appendix, or purchased from any of the reagent houses. In this case 100 ml. of the water is poured into a flask, and sufficient N/10 nitric acid is added to neutralize. This amount of acid will be found in determining the temporary hardness of the water (see (b) below). Four drops of the acid are then added in excess. The solution is then boiled for three or four minutes to expel carbon dioxide, cooled, and about 0·5 ml. of phenolphthalein indicator added. The solution is then titrated with N/10 caustic soda until a very faint pink colour reappears. This colour is then destroyed by the addition of one drop of N/10 acid. The standard potassium palmitate is then run into the solution, in drops while shaking, until a faint pink colour is obtained. If doubtful about this colour note the burette reading and add another two drops of palmitate. A very distinct pink colour should develop if the first reading was correct. When the faint pink colour is obtained the reaction between the hardness and the palmitate is then complete and the amount of palmitate used is a measure of the hardness. The burette reading multiplied by five gives the hardness expressed as parts of calcium carbonate ($CaCO_3$) per 100,000.

Another method which is sometimes adopted for cases where the total hardness may be very high is as follows. A few drops of concentrated hydrochloric acid are added to 210 ml. of the water and the whole evaporated to about 100 ml. A few drops of methyl orange indicator are then added and the solution made neutral by running in N/10 caustic soda until the colour just changes. 20 ml. each of N/10 caustic soda boiled for a quarter of an hour. The solution is then filtered, the deposit and filter paper thoroughly washed with hot distilled water, and the and N/10 sodium carbonate solution are then added and the whole filtrate titrated with N/10 sulphuric acid to neutrality, using methyl orange as indicator. Then the hardness in parts $CaCO_3$ per 100,000 is given by (40—ml. of N/10 sulphuric acid used) × 1·162.

(*b*) Temporary Hardness of Raw Water

This is determined by taking 100 ml. of the raw water, adding a few drops of methyl orange, and titrating with N/10 nitric acid until the colour just changes from yellow to pink. Then the number of ml. of acid used multiplied by five gives the hardness in parts $CaCO_3$ per 100,000. If 70 ml. of water had been taken at the start, the burette reading multiplied by five would give the hardness in degrees, or grains per gal.

This figure is also often called the TOTAL ALKALINITY.

(*c*) Permanent Hardness of Raw Water

This is obtained by deducting the temporary hardness from the total hardness, as determined above.

It may also be determined as follows: Two hundred ml. (200 ml.) of the water are boiled in a flask so as to expel most of the carbon dioxide; 50 ml. of a mixture of equal parts of N/10 caustic soda and N/10 sodium carbonate solutions are then added and the whole boiled down to about 80 ml., cooled, made up to 100 ml., and if possible allowed to stand for some hours to settle. Then 50 ml. are filtered off and accurately titrated with N/10 sulphuric acid. If the caustic soda and sodium carbonate solutions are not exactly N/10 it is essential that the amount of N/10 sulphuric acid required to neutralize them should be known exactly. Then if this amount is x ml. and 50 ml. of the filtered water consumed y ml. of N/10 acid, x-y ml. of N/10 acid corresponds to the permanent hardness in 100 ml. of the original water, and the hardness in parts calcium carbonate per 100,000 is obtained by multiplying this difference by five.

(*d*) Magnesia Hardness of Raw Water

The magnesium hardness of a water may be due to any magnesium salt. Magnesium bicarbonate is included in the temporary hardness, while the other magnesium salts are included in the permanent hardness.

The determination of magnesium hardness is made after the temporary hardness is neutralized by N/10 sulphuric acid and the carbon dioxide boiled off. From this point there are two methods of procedure, depending upon whether lime-water or potassium palmitate is used.

LIME-WATER WORKING DETAILS:—100 ml. of the water, neutralized with N/10 sulphuric acid in presence of methyl orange (that portion used in determining temporary hardness will do) is boiled down until all the carbon dioxide is expelled and the bulk reduced below 100 ml. It is then mixed in a 200 ml. flask with 50 ml. of filtered lime-water, not more than half saturated, the equivalent of which in N/10 acid is accurately known. The whole is raised to 100° C. on a water bath, corked or stoppered and allowed to cool, made up to the 200 ml. mark with well-boiled distilled water, thoroughly mixed and allowed to stand until settled. It is then filtered rapidly while covered to prevent absorption of carbon dioxide, or preferably 100 ml. is pipetted off clear and titrated with N/10 sulphuric acid with phenolphthalein as an indicator. Then if 50 ml. of the lime-water require x ml. of N/10 acid and 100 ml. of the mixture require y ml. of N/10 acid, $x—y$ ml. N/10 acid correspond to the magnesia hardness in 50 ml. of the original water. Each ml. of acid corresponds to magnesia hardness equivalent to 5 parts of calcium carbonate per 100,000.

POTASSIUM PALMITATE WORKING DETAILS:—Take 200 ml. of the water and neutralize with N/10 nitric acid (amount required found in (b) above), plus four drops of acid to make the solution definitely acid. Boil for two or three minutes to expel carbon dioxide. To the boiling solution add 5 ml. of saturated sodium oxalate solution and continue boiling for one minute. This throws down the calcium and leaves a residual hardness due entirely to magnesium. Cool and add a few drops of phenolphthalein indicator. Neutralize with caustic soda and acid as described before and run in potassium palmitate solution, shaking thoroughly, until a faint pink colour appears. Divide the burette reading by 2. Then each millilitre is equivalent to 5 parts of $CaCO_3$ per 100,000.

(e) Hardness Caused by Calcium Salts

This may be found by subtracting the magnesia hardness from the total hardness.

CHEMICALS REQUIRED FOR TREATMENT

The reactions underlying the *lime-soda method of softening* water have already been described, but it will be convenient to recapitulate briefly the essential points. Temporary hardness is precipitated by the addition of hydrated lime, while soda ash (sodium carbonate) precipitates permanent hardness due to calcium salts. Representative equations are as follows:

TEMPORARY HARDNESS.

$Ca(HCO_3)_2$	+	$Ca(OH)_2$	=	$2CaCO_3$	+	$2H_2O$
Calcium bicarbonate		Hydrated lime		Calcium carbonate (deposited)		Water

$Mg(HCO_3)_2$	+	$2Ca(OH)_2$	=	$Mg(OH)_2$	+	$2CaCO_3$		$2H_2O$
Magnesium bicarbonate		Hydrated lime		Magnesium hydroxides (deposited)		Calcium carbonate		Water

PERMANENT HARDNESS.

$CaSO_4$	+	Na_2CO_3	=	$CaCO_3$	+	Na_2SO_4
Calcium sulphate		Soda ash		Calcium carbonate (deposited)		Sodium sulphate (soluble)

$Mg(SO_4)$	+	$Ca(OH)_2$	=	$Mg(OH)_2$	+	$CaSO_4$
Magnesium sulphate		Hydrated lime		Magnesium hydroxide (deposited)		Calcium sulphate

It will be observed that both reagents are necessary for the removal of permanent hardness caused by magnesium salts. Therefore in calculating the amounts of chemicals required for softening it must be remembered that the temporary magnesia hardness requires twice as much lime for its removal as temporary hardness caused by calcium. This means that if we take the sum of the temporary hardness and total magnesium hardness, then the temporary hardness due to magnesium will be included twice, and so the lime required can be quite simply calculated directly from this sum. Again, while permanent hardness due to magnesium is removed by lime, an equivalent amount of calcium permanent hardness is formed. This means that the soda ash required can be calculated directly from the total permanent hardness.

The actual figuring is quite straightforward if all hardness is expressed as parts of calcium carbonate per 100,000 since

1 part $CaCO_3$ per 100,000 requires 0·74 parts of hydrated lime ($Ca(OH)_2$) to remove total temporary hardness and magnesia permanent hardness.

1 part $CaCO_3$ per 100,000 requires 1·06 parts of soda ash (Na_2CO_3) per 100,000 to remove permanent hardness due to calcium salts.

Now 1 part per 100,000 is equivalent to $\frac{0 \cdot 7}{34.75}$ oz. per gal. Therefore the amount of lime and soda ash corresponding to the above can be readily calculated as 1·18 oz. hydrated lime per 1,000 gal. and 1·69 oz. soda ash per 1,000 gal. for each 1 part of calcium carbonate hardness per 100,000. The general formulae for the quantities of chemicals may therefore be written as follows:

Hydrated Lime. (Temporary hardness plus total magnesium hardness) $\times$ 1·18 oz. per 1,000 gal.

If the quicklime figure is required, the hardness should be multiplied by 0·9 oz. per 1,000 gal.

Soda Ash. (Permanent hardness) $\times$ 1·69 oz. per 1,000 gal.

These calculations are for pure 100 per cent chemicals. If lime, say of 90 per cent $Ca(OH)_2$ content is used, then the hydrated lime figure must be divided by 0·90 to give the correct amount for that quality of lime. Similarly for soda ash, which is usually purchased as 58 per cent alkali, or ash containing 58 per cent sodium oxide, i.e., 98 per cent sodium carbonate.

The above figures are, of course, theoretical, and it is often necessary to use an excess to ensure softening reactions proceeding as completely as possible. The excess desirable for complete softening will vary with individual feed conditions as pointed out elsewhere, but in general may

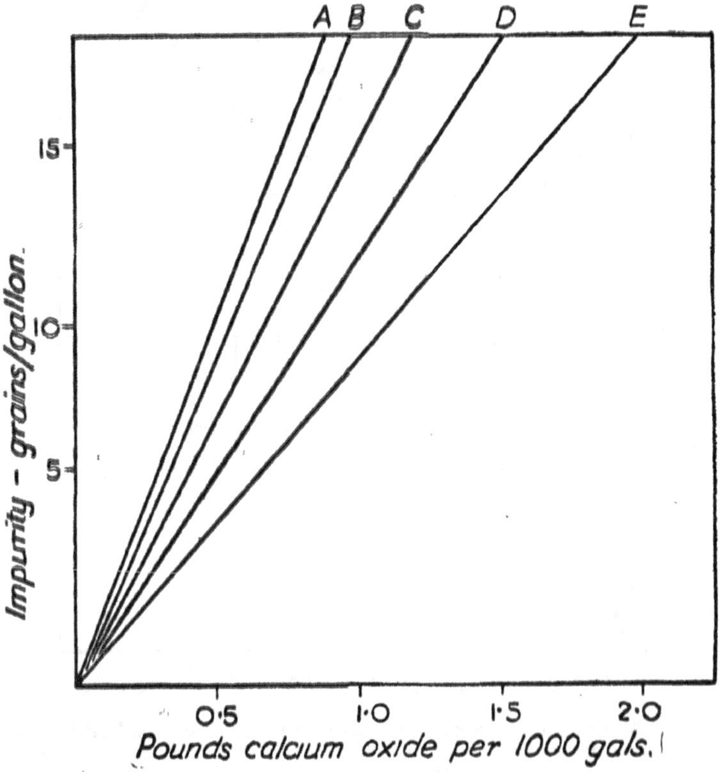

FIG. 35. Amounts of Calcium Oxide required by various salts dissolved in water.

A Calcium bicarbonate.
B Magnesium nitrate.
C Magnesium sulphate.
D Magnesium chloride.
E Magnesium bicarbonate.

The graph may be read for parts per 100,000 of impurity by dividing grains per gallon by 0·7. Calcium oxide may be converted to hydrate of lime by multiplying by 1·32.

be taken as about 1 oz. over the theoretical quantities. In any case, modifications in quantities may be necessary, and these should be based on control tests carried out on the treated water. A further point is that in addition to the above lime, lime equivalent to the free carbon dioxide must be added. If free carbon dioxide has not been estimated, about 2 oz. of lime per 1,000 gal. may be allowed for this on an average. If carbon dioxide has been determined, then the lime equivalent may be calculated from the fact that 44 parts of carbon dioxide require 74 parts of pure calcium hydrate.

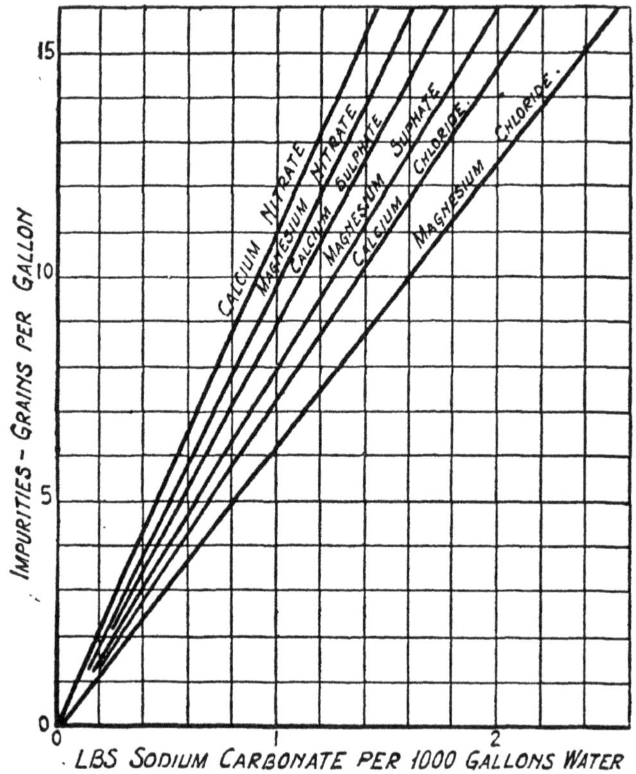

FIG. 36. Amounts of soda ash required by various salts dissolved in water. The graph may be read for parts per 100,000 of the impurities by dividing grains per gallon by 0·7.

From the equations given above it will be observed that *softening with lime alone* can be adopted in certain cases, viz., for removal of temporary hardness and magnesia hardness only. It should be emphasized, however, that any magnesium permanent hardness is not removed but only replaced by an equivalent amount of permanent hardness due to calcium salts. Treatment with lime alone is sufficient where the total alkalinity exceeds the total hardness since in such cases the excess alkalinity is usually due to presence of sodium bicarbonate or carbonate. Water of this type may be obtained from a deep well. Lime only is therefore required for treatment according to the general formula:

Hydrated Lime. (Temporary hardness plus total magnesium hardness) × 1·18 oz. per 1,000 gal.

This figure also is subject to correction as described above.

If some of the alkalinity in the above case were due to sodium bicarbonate this would be converted to an equivalent amount of normal sodium carbonate. This might then give too much alkalinity in the boiler feed and it would be advisable to destroy some of the excess alkalinity by means of calcium chloride. This would throw down calcium carbonate and leave sodium chloride in solution. A similar conversion could be effected with calcium sulphate. The quantities of these chemicals required for such a purpose are given by the following general formulae:

Calcium chloride. (Total alkalinity—Total hardness) × 1·62 oz. per 1,000 gal.

This is for pure 100 per cent chloride; if the hexahydrate is used 3·5 oz. will be required.

Calcium sulphate. (Total alkalinity—Total hardness) × 2·70 oz. per 1,000 gal.

This is for gypsum hydrated with two molecules of water, i.e., $CaSO_4.2H_2O$.

The hardness and alkalinity figures in these formulae are those of the water *after* treatment with lime, and are expressed in parts calcium carbonate per 100,000.

CONTROL TESTS ON TREATED WATER

If condensate is used these tests should be conducted on the feed with condensate, since condensate sometimes shows a slight acid reaction, in which case the alkalinity of the make-up will have to be regulated to take care of this acidity.

(*a*) Total Hardness

This is determined by the methods already described for raw water. Call this H.

(*b*) Alkalinity

The alkalinity of softened water may be due to sodium hydroxide (caustic soda), calcium hydroxide, sodium carbonate, and sodium bicarbonate. A method of testing must therefore be adopted such that any or all of these can be separately determined. This is done as follows: Take 100 ml. of the water to be tested in a beaker and add a few drops of phenolphthalein indicator. If a pink colour is produced this indicates that the water is alkaline to phenolphthalein. (If no pink colour is produced the titration may be conducted with methyl orange indicator as below). Slowly add N/10 sulphuric acid from a burette, gently shaking the beaker and contents, until the pink colour just disappears. Then the number of millilitres of acid used, multiplied by 5, gives P units of alkalinity expressed as parts of calcium carbonate per 100,000. A few drops of methyl orange indicator are then added and more acid run in until the colour of the methyl orange just changes from yellow to red.

The total quantity of acid used right from the start, multiplied by 5, gives M units of alkalinity expressed as parts of calcium carbonate per 100,000. This is the total alkalinity or temporary hardness. P is known as the phenolphthalein alkalinity. This gives all the information required, and the results are interpreted as follows:

(1) If no colouration is produced with phenolphthalein (P=nil) all the alkalinity is present as bicarbonate.

(2) If P is less than $M/2$ the alkalinity is due to carbonate and bicarbonate. The alkalinity as carbonate is $2P$, while that due to bicarbonate is $M-2P$.

(3) If P is greater than $M/2$ the alkalinity is due to hydroxide (caustic alkalinity) and carbonate. The alkalinity as carbonate is $2(M-P)$, while the caustic alkalinity is $M-2(M-P)$.

(4) If P equals $M/2$ the alkalinity is completely caused by carbonate, which therefore equals $2P$.

(5) If $P=M$ the alkalinity is due entirely to hydroxide (caustic alkalinity).

All these alkalinities, of course, are expressed as parts of calcium carbonate per 100,000.

(c) Application of these Results

While the amount of chemicals required for treatment of a feed water are initially determined as outlined under raw water tests, these quantities are approximate and will probably need variation according to the efficiency with which the softening plant is working, e.g., some excess may be required to give complete softening to the limits of the plant. Such variation in quantities is determined by the above control tests of the softened water.

The above tests give three figures: H—the total hardness, P—the phenolphthalein alkalinity, and M—the methyl orange or total alkalinity (which is equal to the temporary hardness). It is often considered that in a properly softened water the ideal is indicated by (1) H as low as possible, (2) M slightly higher than H (by about 1 part $CaCO_3$ per 100,000), (3) P slightly higher than $M/2$ (by about $\frac{1}{2}$ part $CaCO_3$ per 100,000). Special conditions, as pointed out above and referred to more fully later, may necessitate the abandonment of these ideals and call for quite different results in the treated water. For the present, however, these ideals may be examined in somewhat more detail.

Sometimes objection is raised to reduction of the *hardness as low as possible,* since such soft waters are said to be more corrosive. This is quite a mistaken idea and has probably arisen owing to the danger caused by unsoftened natural waters which contain small amounts of corrosive salts, as described earlier in this book. It is essential, of course, that a soft water should fulfil the other conditions of efficient softening, such as alkalinity, and not be too highly concentrated in sodium salts.

If *M is less than H*, a deficiency of soda is indicated, whilst if it exceeds the hardness by more than about 1 part per 100,000 (as $CaCO_3$), then too much soda ash is being used. From the figures already given under softening of raw water, it will be observed that each unit of hardness requires 1·69 oz. of pure soda ash per 1,000 gal. Therefore the proportion of soda ash must be increased or decreased by this amount for every unit of difference between the alkalinity M and the hardness H until the desired conditions are obtained.

If *P is less than M/2*, insufficient lime is being used. It is desirable that P should be rather more than $M/2$ since then there is caustic alkalinity present. This promotes efficient removal of magnesium salts, since magnesium hydroxide is insoluble. The amount of caustic alkalinity which is required for optimum removal of magnesia will vary with raw feed conditions, etc., but as a general rule a value for P greater than $M/2$ by about $\frac{1}{4}$ part of $CaCO_3$ will be sufficient. If P is much greater than $M/2$, too much lime is being used. It has already been shown that each part of $CaCO_3$ per 100,000 is equivalent to 1·18 oz. pure hydrated lime per 1,000 gal. The proportion of lime must therefore be increased or decreased by this amount for a difference of each such unit of alkalinity as indicated by the test.

Finally, if alteration of the proportions of both lime and soda is necessary, it will usually be indicated by

(a) H high, M less than H, and P much less than $M/2$.
In this case more lime and soda is necessary.

(b) H low, M too high, and P too high.
This means less lime and soda is required.

In case (b), however, individual plant conditions may render this departure from the foregoing lime-soda ideals quite satisfactory, or even imperative.

If carbonate conditioning is resorted to in order to prevent scale deposition after softening in the lime-soda plant then it will be essential that the total alkalinity M should exceed the hardness H by an amount sufficient to maintain the required CO_3/SO_4 ratio at the operating boiler pressure. This would probably mean a difference much greater than that recommended above. A consequence of this, of course, is that more lime must be used to keep P above $M/2$ and ensure precipitation of magnesium as hydroxide. It should be remembered, also, that the sodium carbonate is gradually converted into caustic soda in the boiler and it is therefore not sufficient to maintain the ratio at or just above that required for the boiler pressure. The actual ratio in the feed water should therefore be sufficiently high to ensure that the ratio in the boiler water will not fall below the critical value at the maximum concentration attained.

Sodium Aluminate with Lime-Soda Softening

The advantages of sodium aluminate in conjunction with lime and soda have already been enumerated. The use of ¼ lb. or less of this chemical per 1,000 gal. will usually be sufficient to assist in reducing hardness without a large excess of soda (i.e., particularly for cases where carbonate conditioning is not adopted), and will also assist in giving good filtering.

CARBONATE AND PHOSPHATE CONDITIONING

Where carbonate or phosphate conditioning of boiler water is in operation, regular testing of the boiler water to ascertain the carbonate/sulphate or phosphate/sulphate ratio is necessary to ensure that correct conditions are being maintained.

The chemical equivalents of the pure reagent, i.e., no water of crystallization or other impurity—100 per cent phosphate reagent, are as follows:

Sodium metaphosphate ..	0·097 lb. per 1,000 gallons per grain per gallon hardness, expressed as $CaCO_3$.
Trisodium phosphate ..	0·156 lb. per 1,000 gallons per grain per gallon hardness, expressed as $CaCO_3$.
Disodium phosphate ..	0·136 lb. per 1,000 gallons per grain per gallon hardness, expressed as $CaCO_3$.
Monosodium phosphate ..	0·114 lb. per 1,000 gallons per grain per gallon hardness, expressed as $CaCO_3$.

These amounts should be suitably increased, of course, to allow for any water of crystallization or other cause of lack of strength of the reagent. In conditioning the reagent should be used in amount equivalent to the residual hardness in the water plus the excess required to maintain a reserve in the boiler water for dealing with any extra calcium accidentally admitted.

Sulphate Radical.—This may be estimated by the usual analytical method, i.e., by addition of barium chloride solution to 100 ml. of the boiler water sample which has been made slightly acid with hydrochloric acid. The sulphate so precipitated is barium sulphate, and should be filtered, washed with distilled water, ignited, and weighed. Then SO_4 = weight of barium sulphate in grams × 411·5 parts per 100,000.

Sulphate may also be determined volumetrically as follows. Take 20 ml. of the filtered boiler water and add it to 10 ml. of benzidine solution (preparation in the Appendix). Dilute with 30 ml. of distilled water. Allow to stand for 15 minutes and then filter cold through a No. 4 Whatman filter paper. Wash free from acid with as little distilled water as possible; use about 3–5 ml. of water for each washing. Transfer the washed precipitate and filter paper into a suitable conical flask, add 50 ml. distilled water and warm to 50° C. Add a few drops of phenolphthalein indicator solution, and titrate with N/10 caustic soda solution until a pink colour is just obtained. Heat the solution to the boil and then titrate again with a drop or two of the N/10 caustic soda until the pink colour is recovered. Then the amount of alkali used in ml. × 24 gives parts SO_4 per 100,000.

Another method which is quick but only approximate consists in comparing the opacity to light of a known volume of an acidified sample of boiler water to which barium chloride has been added, with that obtained with a standard sulphate solution.

Carbonate Radical.—This can be determined from the alkalinity readings of the boiler water. If the total number of millilitres of N/10 acid used with methyl orange and phenolphthalein is M, and the number to phenolphthalein only is P, then the CO_3 radical, in parts per 100,000, is given by $1 \cdot 26\ (M—P)$. Some operators prefer not to determine carbonate radical in this way, but control the conditioning as follows. The sulphate contents of feed water and blow-down are determined and the ratio (sulphate in blow-down)/(sulphate in feed) determined. If this ratio agrees with the ratio (chloride in blow-down)/(chloride in feed) (see blow-down tests for chloride determination) then it is considered that all the sulphate in the feed water is kept in solution and there has been no deposition of sulphate scale. If, however, a sulphate ratio is less than the chloride ratio it is said that sulphate is being deposited as scale, thus calling for increased soda ash feed. A greater sulphate ratio indicates that old sulphate scale is being dissolved and feed conditions should be maintained until it is found that sulphate is being deposited.

Phosphate Radical.—The amount of phosphate in the boiler water is usually found by colorimetric methods specially developed for testing in the presence of silica. Older methods gave incorrect readings in the presence of silica or silicates. The solutions required are: (1) A standard phosphate solution containing 100 p.p.m. of PO_4 is obtained by dissolving 0·1390 grams of dibasic ammonium phosphate in distilled water, and making the solution up to 1 litre. The analytic reagent (A.R.) quality of reagent should be used. (2) Sulphuric acid solution—65 ml. of concentrated A.R. acid mixed with distilled water and the solution made up to 500 ml. (3) Ammonium molybdate—17·5 grams of the A.R. reagent is dissolved in about 200 ml. of distilled water. To another 200 ml. of distilled water add 10·5 ml. of concentrated A.R. sulphuric acid. Mix these two solutions, make the volume up to 500 ml. and store in the dark. (4) Hydroquinone—*dissolve* 5 grams of pure hydroquinone in 500 ml. of distilled water and add 0·3 ml. of conc. A.R. sulphuric acid after solution is complete. Store the solution in the dark to retard development of the brown coloration. New solution will be required about every three weeks, since the increasing darkening makes colour comparisons difficult. (5) Alkaline sulphite solution—in 500 ml. distilled water dissolve 100 grams anhydrous sodium carbonate and 24 grams sodium sulphite, both A.R. quality.

WORKING DETAILS.—Take 5 ml. of the filtered boiler water in a 50 ml. Nessler tube. Using separate pipettes for measuring, add in the following order 1 ml. sulphuric acid solution, 1 ml. molybdate solution, and 1 ml. hydroquinone solution. During each addition, and in between additions, shake the tube and mix thoroughly. Now allow to stand 5 min. While this mixture is standing, measure 2 ml. of the alkaline sulphite solution into another Nessler tube. At the end of the 5 min. pour the mixed solution into the alkaline sulphite; mix thoroughly, by some such means as pouring back and forth several times. Phosphate will develop a blue colour. This should be compared with standards at once, as the phosphate colour starts to fade quite soon.

Standards for colour comparison can be prepared for each test by taking known amounts, say 1, 2, 3, 4, 5 ml. of the standard solution in 5 Nessler tubes and making each up to 5 ml. Each of these standards would then be treated in exactly the same way as the 5 ml. of boiler water. These standards would then have colours for 20, 40, 60, 80, and 100 p.p.m. of PO_4. The test sample would match one of these or be in between two of them. If it is known that the PO_4 content is about, say, 30 p.p.m., only the two standards either side could be prepared, and comparison would then reveal whether the phosphate was within the range.

For efficient testing it is essential that the boiler water sample should be perfectly clear. If colour is troublesome after filtering, it may be removed with carbon, but in this case it is essential to test the carbon. Some carbons give high PO_4 tests and are therefore to be avoided.

To test a carbon sample, use a similar amount to that which will be used for decolorizing the water. Add the carbon to 35 ml. of distilled water and a little dilute caustic soda. Allow to stand about the time which would be taken in treating with carbon. Filter, and test the filtrate for PO_4 as above.

DISSOLVED OXYGEN

The chemical method most commonly used for the determination of dissolved oxygen was originated by Winkler in 1888. It is based on the oxidation of manganous hydroxide to manganic hydroxide in the presence of alkali by means of the oxygen dissolved in a measured amount of water. The manganic hydroxide, in the presence of hydrochloric acid subsequently added, reacts with potassium iodide to liberate free iodine. This iodine can be readily titrated with a standard solution of sodium thiosulphate, and as the iodine liberated is equivalent to the dissolved oxygen, the amount of standard thiosulphate required is a measure of the dissolved oxygen content.

SOLUTIONS REQUIRED.—(1) Manganous chloride—40 grams of pure crystallized manganous chloride are dissolved in 100 ml. of distilled water. The solution should be free from iron. (2) Alkaline iodide solution —360 grams of caustic soda and 100 grams of potassium iodide are dissolved in distilled water to give 1 litre of solution. The caustic soda

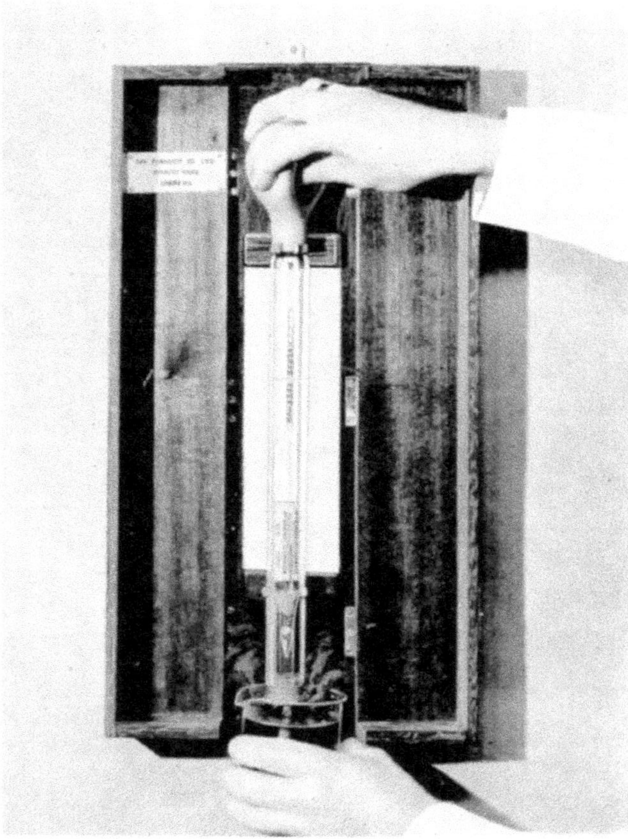

PLATE 25. HYDROMETER WITH SELF-CONTAINED TEMPERATURE CORRECTION SCALE
(*Permutit Co., Ltd.*)

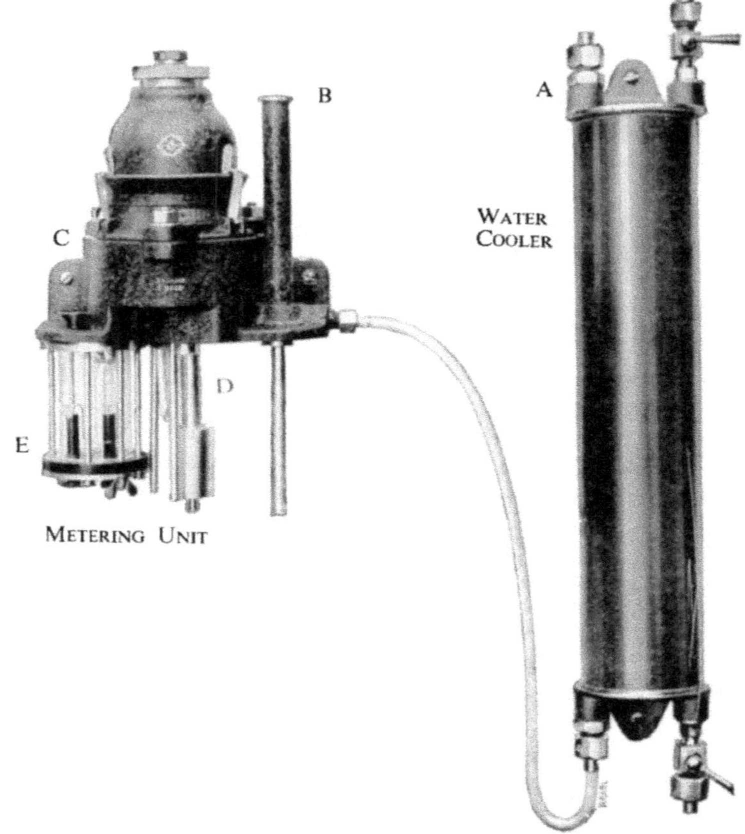

PLATE 26. CAMBRIDGE DISSOLVED OXYGEN RECORDER
(*The Cambridge Scientific Instrument Co., Ltd.*)

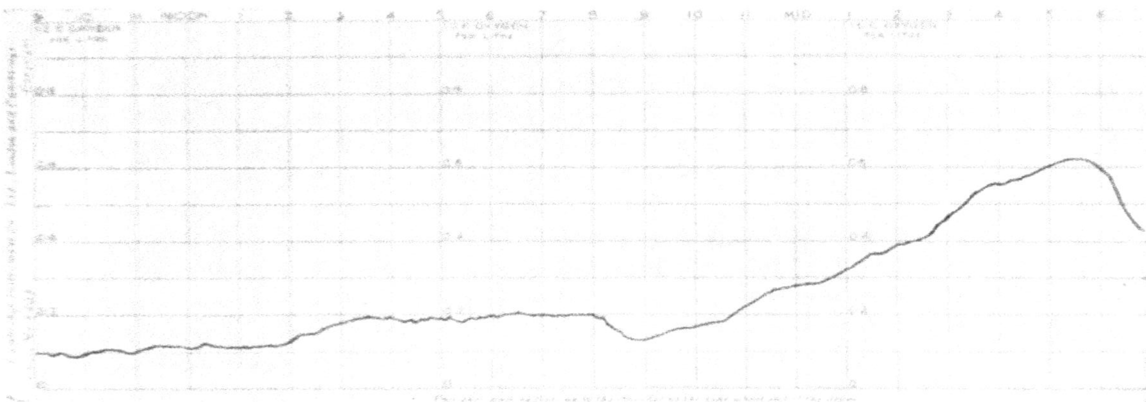

PLATE 27. **REPRODUCTION OF AN ACTUAL DISSOLVED OXYGEN RECORD**

used must be free from nitrate and should contain only very little sodium carbonate. (3) Hydrochloric acid—usual concentrated solution. (4) Sodium thiosulphate—standard solution, strength one hundredth normal (N/100). This can be made by dissolving 2·48 grams of the pure crystals in distilled water and making up to 1,000 ml. It is advisable to have this made up by a chemist and to obtain a fresh supply once a month owing to its instability in weak solution. (5) Starch Indicator—add 200 ml. boiling distilled water to 4 grams of starch, allow to settle, and decant the clear liquid, which is then ready for use.

A suitable apparatus is a stout glass flask or bottle fitted up after the manner illustrated in Fig. 37, with a capacity of about 250 ml. (the volume of water must be known accurately). A sample of the water to be tested is drawn through A, manipulations such as inverting the bottle, opening and closing the cock B, etc., being adopted to prevent any bubbles of air being left in the bottle. After about 10 minutes' flow of water (a total through the flask of about 25 litres), the bottle should be ready for testing all cocks being closed.

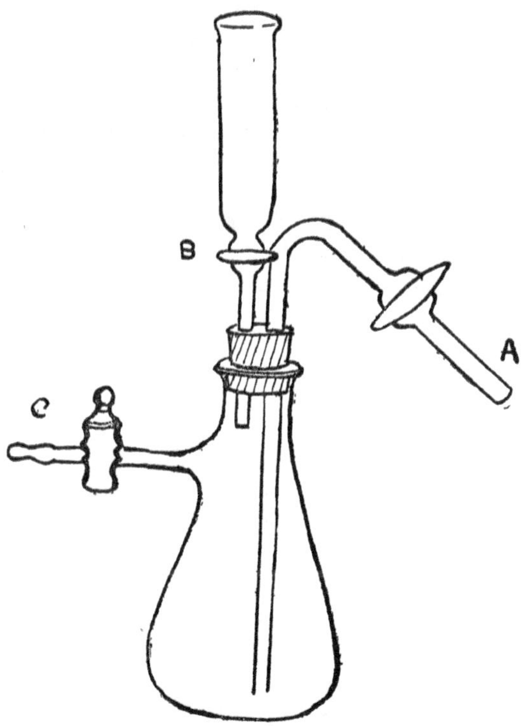

FIG. 37. Apparatus for dissolved oxygen determination.

The method of testing is as follows: 2 ml. each of the manganous chloride and the alkaline iodide are introduced into the flask via the funnel and cock B. The reagents can be admitted without admitting any air by placing in the funnel, opening B, and then cautiously opening C until the reagent has been drawn in. These two reagents should be admitted separately, and the funnel washed out each time with distilled water before fresh reagent is added to it. The contents are then mixed by gentle shaking and turning the bottle. A flocculent precipitate will form which soon settles to the bottom if the shaking has not been sufficiently violent to break it up into a small powder difficult to settle. The clear solution above the precipitate should be almost colourless; if it is brownish the shaking should be repeated. When the solution is clear 5 ml. of fuming hydrochloric acid is placed in the funnel, and the reagent is drawn in as in the case of the others. Before the last drop enters the flask the tap should be closed and the contents shaken again for about 5 min. or so. The precipitate dissolves and leaves the liquid coloured by iodine, and the density of this straw coloration affords a rough indication of the extent of dissolved oxygen present in the feed water. At this stage the liquid may be transferred to a flask and allowed to cool to atmospheric temperature. The standard N/100 sodium thiosulphate solution is then run in until most of the iodine has been decolorized; a few ml. of starch solution is then added and an intense blue compound is formed. Further thiosulphate is then added until this colour disappears. The total volume of thiosulphate taken to reach this point is noted, and since 1 ml. of N/100 thiosulphate is equivalent to 0·056 ml. of oxygen at normal temperature and pressure the volume of oxygen in the amount of water taken, and therefore in each litre of feed water, can easily be found by simple calculation.

Accurate results can be obtained in this way, but it should be noted that if the water contains much organic matter this may react with the acid iodide and so give incorrect results. This may be corrected for by taking equal volumes of distilled water and the water to be tested, adding 10 ml. of N/100 iodine solution, and titrating with N/100 thiosulphate. The difference is then due to organic matter and may be applied as a correction.

Associated with the question of accuracy, of course, is the accuracy

FIG. 38. "Admiralty Standard" dissolved oxygen testing outfit.

of sampling. If an average sample is not obtained, variable results will be given by the test. It should be appreciated that at certain points, such as at pump and valve glands, an increase in oxygen content may occur and sampling points should therefore be located at various points with a view to tracing any cause of faulty operation. A good final sampling point is on the pressure side of the boiler feed pumps.

Another method which is sometimes used is absorption of the oxygen in ferrous sulphate solution This is carried out as follows. The reagents required are (1) Methylene blue indicator (1 part in 1,000) or Phenosafranine indicator (1 part in 2,000). (2) Standard ferrous sulphate solution—0.23 gram of the chemically pure salt ($FeSO_4$) with 1 ml. of conc. sulphuric acid in 100 ml. (3) Alkaline tartrate —dissolve 173 grams Rochelle salt and 52 grams NaOH in 500 ml. distilled water. The method of procedure is to draw 50 ml. of the water to be tested from the feed system, add 5 ml. of the alkaline tartrate solution and 1 drop of one of the indicators, and place the mixture in a Nessler cylinder. The ferrous sulphate solution is then rapidly titrated into the Nessler cylinder until the colour disappears. Each ml. of ferrous sulphate solution required is equivalent to an oxygen content in the feed of 1 ml. per litre.

A useful portable testing set for determination of dissolved oxygen by the Winkler method is the "Admiralty standard set" supplied by Messrs. Griffin and Tatlock, Ltd., and illustrated herewith. The whole outfit is contained in two teak boxes, the larger one for reagents and general equipment, and the smaller box containing a special sampling and analysing bottle.

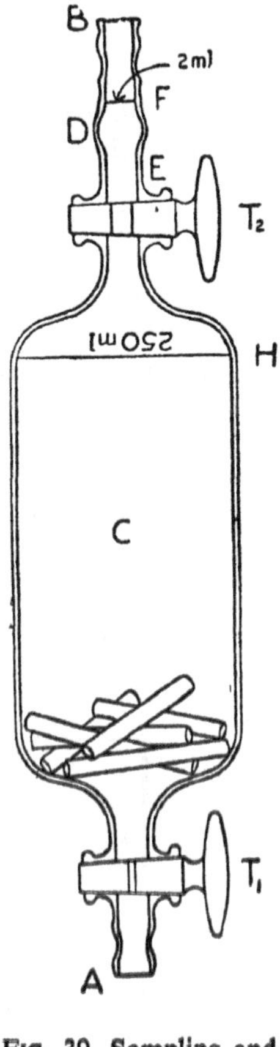

FIG. 39. Sampling and Analysing Bulb of "Admiralty Standard" set.

This sampling and analysing device is illustrated in Fig. 39, and consists of a glass bulb C with an inlet A and an outlet B via stopcocks T_2 and T_1 and is also provided with a small bulb of 2 ml. capacity immediately before the inlet cock.

Sampling and analysis is carried out as follows:

If the solution being sampled is at excess pressure up to 20 to 30 lb. per sq. in., connection is made to A and overflow is carried off from B. The solution is allowed to flow gently through the bottle with both its taps open. When at least six times the volume of the bottle has passed through it, close the sampling cock and then taps T_1 and T_2 in succession.

If the solution is at, or lower than, atmospheric pressure, the bottle is first filled with water from the feed tank. A is connected to the sampling cock at higher pressure and B to the sampling cock at lower pressure. Open B partially, then T_2 and T_1. The sampling cock connected to A is then gradually opened and sample collected as above. If only one sampling cock is available, the solution must be drawn through the bottle by suction applied at B.

Estimation of oxygen is then made as follows: 2 ml. of solution No. 1 manganese chloride is measured into bulb D which carries a mark at 2 ml. Open T_2 and then T_1 so that the solution is sucked into bottle and an equal volume discharged at A. Close T_2. Clean out tube E and bulb D with water. Admit 2 ml. of solution No. 2 alkaline iodide in the same way. Close T_1 and T_2 and agitate gently. Admit 2 ml. of solution No. 3 sulphuric acid through T_2 by gravity, T_1 remaining closed. Gently agitate the bottle.

After a minute or two the stopcock T_2 should be closed and the bottle slowly inverted. If on re-inverting the solution is not clear, open T_2 and agitate until it is so. The bottle is then inverted and liquid run out until the mark H (250 ml.) is reached. The liquid thus run into a conical flask is stirred and a few drops of starch solution added. The thiosulphate solutions are provided for titration of this according to the extent of aeration of the original sample of water. In one of these solutions each ml. used to remove the blue coloration given by starch represents 1 ml. of oxygen per litre; with the other the equivalent is 0·1 ml. of oxygen per litre, and this solution should therefore be used to test the effluent from an efficient de-aerator.

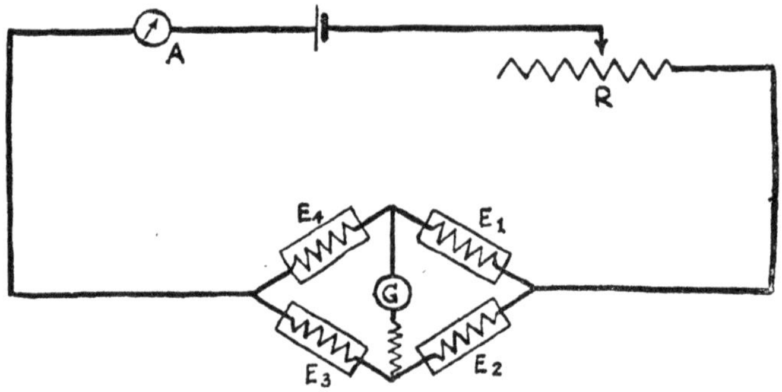

FIG. 40. Wheatstone Bridge Arrangement of Cambridge Dissolved Oxygen Recorder.

An outfit which measures dissolved oxygen by physical means is the dissolved oxygen recorder of the Cambridge Scientific Instrument Co. This gives a continuous record of the oxygen in the feed water and is therefore not subject to the disadvantage of spot chemical tests, viz., that a leak or other cause of increased oxygen content may develop and continue for several hours before detection. The principle of the equipment consists in scrubbing the feed water with pure hydrogen gas so as to displace oxygen and nitrogen by dissolved hydrogen. The exit mixture of hydrogen, oxygen, and nitrogen then passes over one half of a Wheatstone bridge device (Fig. 40) formed of two electrically heated wire spirals (E_1 and E_2), the other half being another two heated spirals (E_3 and E_4). Owing to the difference in the thermal conductivity of nitrogen, oxygen, and hydrogen the heat radiated from each pair of spirals is different, and accordingly a current is generated and causes a deflection in a galvanometer (G). This deflection is proportional to the amount of oxygen and nitrogen, and since the amounts of these two gases dissolved in water usually bear a constant proportion to one another, the deflection can be directly correlated with the dissolved oxygen content.

The metering unit of the apparatus is illustrated at Plate 26. The water to be analysed enters the cooler A from a high-pressure valve. It is here reduced in temperature to about 20–25° C., and then passes on to a constant head device B which maintains the correct flow of water through an orifice situated in a water jacket C surrounding a katharometer. The water then flows down the contact chamber D, where it comes into intimate contact with hydrogen gas, dissolving some of this and giving up part of its oxygen. The hydrogen is generated in the electrolytic cell E, passes the first side of the katharometer and then flows through compensator and restriction to the contact chamber. The second side of the katharometer is exposed to the gas in the chamber. The katharometer consists of four platinum spirals mounted in cells in a central metal block and connected together to form a Wheatstone bridge. Two of the cells are open to the pure hydrogen and two to the gas from the contact chamber. The outfit can be used with either a recorder or an indicator, or a combined indicating and recording outfit can be used. The recorder or indicator can, of course, be mounted some distance away at any convenient point. A reproduction of an actual record is shown at Plate 27.

BLOW-DOWN TESTING

Unless carbonate or phosphate conditioning is in operation, tests on boiler water are usually confined to determinations of total dissolved solids, suspended matter, and alkalinity.

(a) Alkalinity

The alkalinity of the boiler water can be determined in the same way as that of the feed water, or by the barium chloride method described under embrittlement tests.

(b) Dissolved Solids

EVAPORATION METHOD.—From 250 to 1,000 ml. of the water are evaporated in a tared dish (platinum, if possible), adding the water from time to time to make up the bulk as evaporation proceeds. Care is taken to keep the water line at least ½ in. from the top of the dish. The evaporation may be carried out by placing the dish on the top of a spouted beaker, the diameter of which is slightly smaller than that of the dish, and placing a large inverted funnel overhead to prevent any dust from falling in. The dish and residue are finally dried for 2–3 hours in a water oven and rapidly weighed. Care must be taken that the outside of the dish is not contaminated. If platinum is not available a good quality porcelain dish may be used, but glass should be avoided.

This method is rather tedious, however, and quicker and more convenient methods are usually adopted.

DENSITY DETERMINATION.—The amount of solids in solution is roughly proportionate to the density of the solution. The density of the boiler water may therefore be taken as a guide. A useful hydrometer for the purpose is the Twaddell. The test may be made by cooling the sample of boiler water to 60° F. in a glass cylinder, and the hydrometer is then inserted. When it comes to rest the density in degrees Twaddell may be read direct, but it is important that the temperature should be 60° F., or at least very near to it, since density varies with temperature. This method will not give absolutely accurate results owing to the different densities of the different salts in solution. It is, however, a useful approximate guide. The degree Twaddell at 60° F. may be related to the fact that sea water, with about 2,500 grains per gal. dissolved solids, has a density of just over 5 degrees Twaddell. On this basis the degree may be taken as equivalent to 500 grains per gal. Another method of measuring density or saline concentration, of course, is the marine salinometer. This acts just as a hydrometer, and is calibrated in 32nds with water at 200° F. In this case sea water is taken as having a saline concentration of about 2,200 grains per gal. and the marine salinometer reads 1/32nd in such water.

FIG. 41. Twaddell Hydrometer No. 1 (0°–24° or Sp. Grav. 1·00–1·12). Other Twaddells read for higher density ranges, No. 6 reading 138°–170° or Sp. G. 1·69–1·85.

A new type of hydrometer for testing salinity has been recently introduced by Messrs. Permutit Co., Ltd. It has already been mentioned that owing to the change in density of saline solutions with temperature it is necessary to cool the solution to a certain definite temperature or to apply a correction from tables. The new hydrometer has an integral thermometer in the lower part of it, as illustrated (Plate 25), so that the operator is able to read directly the temperature density correction on the instrument. This therefore avoids the use of a separate thermometer or correction tables.

In the illustration the instrument shows a reading as follows:

Stem reading	600 g.p.g. sodium chloride
Lower scale reading, plus	100 g.p.g. „ „
Correct concentration	700 g.p.g. „ „

The accuracy of reading is 20 grains per gal.

Chloride Ratio Measurement.—A very convenient and satisfactory method of watching concentration in the boiler is to determine the chloride in the feed and in the blow-down, and to maintain the ratio of the two figures as constant as possible by blowing down. Chloride normally passes through the boiler unchanged and therefore permits the use of this ratio as a direct measure of the concentration of the water in the boiler.

The chloride in the feed water may be determined by taking 300 ml. of the water and evaporating down to about 100 ml., adding 1 drop of methyl orange indicator and then running in N/10 nitric acid from a burette very slowly until the solution is just faintly pink. Add 2 drops of potassium chromate indicator solution and run in N/10 silver nitrate from a burette very slowly with constant stirring. A white precipitate forms, with a brick-red coloration which disappears on stirring. Silver nitrate is added until the brick-red coloration is permanent. A blank determination should be made using distilled water instead of the sample. Then the difference in the amounts of silver nitrate used in testing sample and blank is equivalent to the chloride present. The number of ml. divided by 3 and multiplied by 3·55 gives the number of parts of chlorine (Cl) per 100,000. If multiplied by 5·85 instead of 3·55, the parts of sodium chloride per 100,000 will be obtained, but this assumes, of course, that all the chlorides present are as sodium chloride.

The chloride in the boiler water can be estimated in precisely the same way, but using only 20 ml. of the filtered blow-down. Then the number of ml. of silver nitrate required, minus the blank, should be multiplied by 5 before calculating the amount of chlorine per 100,000 as above.

If the water is strongly alkaline, it should be neutralized with N/10 sulphuric acid before estimating chlorides as above.

(c) Suspended Matter

This is determined by filtering 1,000 ml. of the water (taken after carefully mixing a large quantity) through a small dried and ashless quantitative filter paper. The filter is washed with distilled water, dried at 105–110° C. till constant, and weighed. The gain in weight is then "mud and suspended matter." The filter is now ignited, and the residue is "suspended matter," while the difference between the two last figures gives the "organic suspended matter." All the weights in grams, if divided by 100, will be parts per 100,000.

EMBRITTLEMENT RATIOS

Methods for determining sulphate, chloride, and carbonate have already been described.

The caustic alkalinity can be found as described already during the alkalinity test. An alternative method is as follows:

Take 100 ml. of the boiler blow-down sample and add 25 ml. of 10 per cent barium chloride solution. It may be necessary to dilute the solution at this stage so that a pink coloration is produced on adding a few drops of phenolphthalein. Run in N/10 nitric acid from a burette, a drop at a time, until the pink colour just disappears. Then the number of ml. of acid used, multiplied by 5, gives the parts of caustic alkalinity, expressed as parts $CaCO_3$ per 100,000.

This method may also be used for determining caustic alkalinity in the softened water, but in that case 100 ml. of the water should be mixed with 10 ml. of the barium chloride.

GENERAL WATER ANALYSIS

It is occasionally necessary to carry out a full analysis of a water, such as when the use of a new source of supply is contemplated and it is necessary to be sure that there is nothing harmful present which might not be revealed by the routine tests previously described. Such additional estimations as would be required under these circumstances are briefly set out below.

(*a*) Free Carbonic Acid (Carbon Dioxide)

This is determined by titrating 100 ml. of the water with N/10 sodium carbonate, using phenolphthalein as an indicator, until a faint permanent pink coloration is obtained. The soda solution should be added very slowly, and it is usually advisable to carry out a similar estimation on a sample of distilled water which has been well boiled and cooled. The permanent pink colorations obtained in the blank and on the sample should be equal in intensity.

Then each ml. of sodium carbonate solution used for titration corresponds to 2·2 parts of free CO_2 per 100,000 or 3·1 parts of carbonic acid (H_2CO_3).

(*b*) Silica

Acidify 500 ml. of the water with hydrochloric acid (5–10 ml.) and evaporate to dryness. The residue is moistened several times with hydrochloric acid, completely evaporated after each addition, and then dissolved in hot water with addition of hydrochloric acid. The silica, which forms a deposit in this solution, is collected on a small ashless filter paper, washed successively with dilute hydrochloric acid and twice with hot water. The filtrate up to this point may be used for estimation of lime, magnesia, etc. The silica is then washed again several times, the paper and deposit ignited, and the residue weighed as silica.

Then the weight of silica in grams multiplied by 200 gives the number of parts per 100,000.

(*c*) Lime

The filtrate from the estimation of silica is used for this determination, and it may contain iron, alumina, magnesia, and manganese in addition to the lime. These must be removed. Add about 10 ml. of ammonia, cover the beaker of liquid with a watch glass, and heat on a water bath (boiling) for about five minutes. If any precipitate forms, continue heating the solution until the precipitate coagulates, when it is filtered through an ashless filter paper and thoroughly washed. This contains iron, alumina, and manganese, and is dealt with as described below. The filtrate (or original solution if there has been no precipitate with ammonia) is heated to boiling, and a slight excess of hot ammonium oxalate solution added drop by drop while stirring or shaking the beaker. Allow the precipitate to settle and the solution to cool somewhat; then add a few drops of ammonium oxalate to make sure all the lime has been thrown out. Filter through an ashless filter paper and wash at least half a dozen times with hot water. Ignite the precipitate and weigh as calcium oxide. To be sure that the precipitate of calcium oxalate has been completely decomposed to lime, the ignition and final weighing should be repeated until a constant final weight is obtained.

Then weight in grams, multiplied by 200, gives the parts of lime (CaO) per 100,000.

(*d*) Magnesia

The filtrate from the lime estimation is heated to boiling and a solution of microcosmic salt ($NaHNH_4\ PO_4+H_2O$) is added until no further precipitation takes place. Most of the magnesium is thrown down in this way. Then add concentrated ammonia equal to about one-third of the volume of the solution, stirring the solution during addition. If possible, allow to stand overnight to allow precipitate to settle, but at least allow two hours; in the latter case, preliminary vigorous shaking for half an hour will assist coagulation and settling. Filter through an ashless paper, wash at least half a dozen times with dilute ammonia, and dry in a steam oven. The ignition of the dry precipitate should be carried out very carefully, ignition being gentle at first until all the ammonia is driven off and then gradually increased. Finally, remove the lid of the crucible to allow any carbon to burn off. Weigh.

Then weight in grams, multiplied by 72·4, gives parts of magnesia (MgO) per 100,000.

(e) Iron

The precipitate obtained with ammonia when estimating lime above contains iron, alumina, and manganese. If the quantity is small it will be sufficient to ignite and weigh and report it as a mixture, when, if multiplied by 200, will be expressed as parts mixed oxides per 100,000.

If it is desired to estimate each constituent separately, the precipitate should be dissolved in hot hydrochloric acid and the solution then poured into excess of a boiling solution of caustic potash or caustic soda. Be sure the resulting mixture is strongly alkaline. Boil until the precipitate coagulates, allow to settle, filter, and wash. This precipitate contains the iron and manganese. (The filtrate contains the alumina.) Ignite the precipitate and weigh as mixed oxides of iron and manganese. Then weight in grams, multiplied by 200, gives parts per 100,000.

Manganese in water is not usually appreciable, so that it need not be separately estimated. If, however, there is reason to suspect that the water contains measurable amounts of this, the iron may be estimated colorimetrically and the difference will then be manganese oxide.

(*f*) Alumina

The filtrate from the iron estimation is acidified with hydrochloric acid, heated to boiling, and strong ammonia added until in very slight excess. Boil for about five minutes. Filter. Wash. Ignite, and weigh as alumina.

Then weight in grams, multiplied by 200, gives the parts per 100,000.

(g) Nitrates

These may be estimated colorimetrically, using phenolsulphonic acid, or by Lunge's nitrometer method. Only the former will be described here, as the latter may be obtained on reference to any standard analytical work of reference.

First add a weighed quantity of silver sulphate to the water to throw out almost all the chlorides. This is assisted by heating. Filter the solution and use the filtrate as the sample.

Evaporate 20 ml. or less of the sample in a small evaporating dish on a water bath, removing it from the water bath just before it becomes dry, and allowing the last few drops to evaporate at room temperature in a place protected from dust. The volume given above depends largely on the amount of nitrate present, and the figure given is as a general guide. If this proves unsuitable at a later stage in the test, the operation will have to be recommenced, using a larger or smaller volume of water. If the water develops acidity on evaporation, a new start will also have to be made; the water should then first be stirred with a little calcium carbonate, evaporated almost to dryness, filtered, and then taken right down to dryness. Add about 2 ml. of the phenolsulphonic acid (prepared as described in the Appendix) to the cool residue, rubbing it quickly and thoroughly over the residue with a glass rod, then dilute with cold distilled water and make alkaline with caustic soda or potash. A yellow colour should then be developed. Transfer the liquid to a 100 ml. Nessler tube and fill to the 100 ml. mark with distilled water. This may be compared with permanent standards kept for the purpose, diluting it in stages until it matches one of the standards exactly. To prepare the standards, dissolve 0·72 gram of pure recrystallized potassium nitrate in 1 litre of distilled water. Take 10 ml. of this solution and evaporate cautiously on a water bath, as in the case of the sample above; moisten the residue quickly and thoroughly with 2 ml. of phenolsulphonic acid and dilute to 100 ml. This will contain 1 part of nitrogen per 100,000. Portions of this can be diluted to give lower standards, and by taking 20, 30, etc., ml. of the nitrate solution to start with stronger standards can be made.

Now if 100 ml. of the water was taken and chlorides removed with silver sulphate, then 20 ml. evaporated and finally taken to give 100 ml. of a coloured solution, this coloration will represent one-fifth of the nitrogen in the water. Suppose, too, that only 25 ml. of this 100 ml. were compared with a standard tube and that it had to be diluted to 40 ml. to match the 1 part nitrogen standard. Then the nitrogen present in this 25 ml. would be $\frac{40}{25} \times 1$ parts per 100,000; therefore the nitrogen present in the complete 100 ml. of coloured solution would be $4 \times \frac{40}{25} \times 1$ parts per 100,000; as this was made up with only one-fifth of the original water sample, we have a nitrogen content of $5 \times 4 \times \frac{40}{25} \times 1$ parts per 100,000 in the water being tested. Nitrogen (N) is converted to nitric anhydride by multiplying by 3·85.

(*h*) Ammonia

Detected by adding a Nessler solution. This is a solution of the double compound of potassium and mercuric iodides (K_2HgI_4) in caustic solution. It forms an extremely sensitive reagent for the detection of ammonia, a strong brownish yellow coloration being obtained even with extremely dilute ammonia solutions. The reaction is so sensitive that 0·05 mg. of ammonia can be detected in 1 litre of water.

Estimation of ammonia, if required, is carried out as follows: The calcium and magnesium compounds present may interfere with the method, so these must first be removed by means of sodium carbonate and caustic soda. Filter off the precipitate. A suitable quantity (say 50 ml.) of the filtrate is then placed in a Nessler tube and the solution diluted up to the 100 ml. mark with distilled water. Then add 2 ml. of Nessler solution and stir. A similar Nessler tube is made up in the same way with distilled water and Nessler solution only. Standard ammonium chloride is then added to the latter in small quantities until, after standing two or three minutes, a colour match with the test sample is obtained. When almost sufficient standard ammonium chloride has been added from the burette the second Nessler tube should be made up with distilled water until its volume is about equal to that of the sample tube. Then the volume of ammonium chloride required, in ml., measures the amount of ammonia in the water. This may be checked in a repeat test, using both Nessler tubes full, since the approximate quantity of ammonium chloride will be known, and the second tube can therefore be filled more nearly to comparison with the sample tube.

As a general rule the quantity of water used for testing should not be such that more than 2 ml. of the standard ammonium chloride solution is required to produce a colour match.

The number of ml. of ammonium chloride solution required, divided by the volume of water taken, gives the parts of NH_3 per 100,000.

(*j*) Oil

At least a litre of the water, taken in small quantities at a time, is shaken with methylated ether in a separating funnel; the ether layer in the funnel should always be at least 1 in deep. Distil off the ether in a distilling flask, and dry the oil to constant weight in a steam oven.

Then the weight in grams, multiplied by 100, gives the oil in parts per 100,000.

(*k*) Calculation of Results

One cannot say with certainty the exact combinations of acids and bases found by analysis, but it will be obvious that some system is essential to give an approximation to their combination. On the basis that in a mixture of several salts the least soluble salts are formed, the following instructions may be followed in calculating the amounts of individual salts. Any variations may be made at the discretion of the analyst:

Iron, aluminium, ammonia, and manganese are usually present as sulphates and as their amounts are not generally large they may be calculated as such.

All the lime, or as much as possible (according to the alkalinity), is then calculated to calcium bicarbonate.

Any remaining lime, or as much as possible, successively to sulphate, nitrate, and chloride.

If any alkalinity remains after calculating calcium bicarbonate then all the magnesia, or as much as possible (depending on the alkalinity), is calculated to magnesium bicarbonate.

Any remaining magnesia, or as much as may be possible, successively to sulphate, nitrate, and chloride.

Any remaining alkalinity (caustic or carbonate) must then be expressed as the sodium compounds, and any remaining radicles quoted as sodium salts. Silica is usually expressed as such.

ANALYSIS OF SCALES

It may sometimes be desirable to determine the composition of a scale from the boilers, or turbines, etc., in which case the following brief scheme of analysis may be followed:

(*a*) Water

Heat 1 gram in an oven at about 250–300° F. until constant in weight.

(*b*) Organic Matter

Heat the residue from the above determination to very faint redness, with the lid off the crucible. The loss in weight will be organic matter, oil, and any combined water not removed above.

(*c*) Oil

Similar methods to those adopted in water analysis.

(*d*) Silica

Take the sample from the organic matter estimation, or burn the organic matter off a fresh sample, add hydrochloric acid, boil to dissolve as much scale as possible, and evaporate to dryness. Bake on a hot plate for about an hour. Cool, add 5–10 ml. hydrochloric acid, heat, and dilute with about 100 ml. of distilled water. Boil. Filter, wash, dry, and ignite the silica.

(*e*) Sulphates

The filtrate from the silica determination is evaporated to about 100 ml., and sulphates precipitated with barium chloride as described under embrittlement ratios. This gives the sulphates on 1 gram of scale.

(*f*) Iron, Aluminium or Manganese

These may be determined together by precipitating with ammonia from the filtrate from sulphate estimation, or they may be separately estimated as described under water analysis. This also is the weight on 1 gram of scale.

(*g*) Lime

Ammonium oxalate is added slowly to the filtrate from the iron, etc., until no further deposit is formed. The deposit is filtered, washed, ignited, and weighed as the lime on 1 gram of scale.

(*h*) Magnesia

Acidify the filtrate from the lime estimation with hydrochloric acid, evaporate to about 100 ml. or rather less, make alkaline with ammonia, and then throw down magnesia with sodium phosphate as in water analysis. The weight of precipitate, which is pyrophosphate, multiplied by 0·362, gives the magnesia on 1 gram of scale.

(*j*) Calculation of Results

All the sulphate to calcium sulphate.

All the magnesia to magnesium hydroxide.

Any lime left after calculating calcium sulphate should be expressed as calcium carbonate.

Silica, although it may be present combined with calcium and magnesium as a complex silicate, is usually expressed separately.

If the total, including organic matter, water, etc., exceeds 100 per cent, a sufficient quantity of the magnesia may be expressed as oxide instead of hydroxide to reduce the figure to 100 per cent.

RESIDUAL CHLORINE

Where chlorination of condenser cooling waters is adopted, the amount of residual chlorine left in the effluent circulating water may be determined as follows:

SOLUTIONS REQUIRED:

(a) *Orthotolidine solution.*—Dissolve 1 gram of orthotolidine, m.p. 129° C., in 1 litre of dilute hydrochloric acid (100 ml. of the concentrated acid diluted to 1 litre). It is also advantageous to facilitate solution of the orthotolidine by first grinding it up in a small mortar.

(b) *Copper sulphate solution.*—Dissolve 1·5 gram of copper sulphate and 1 ml. of concentrated sulphuric acid in distilled water and make up the total volume to 100 ml.

(c) *Potassium dichromate solution.*—Dissolve 0·25 gram of potassium dichromate and 1 ml. of concentrated sulphuric acid in distilled water, making the total volume up to 1 litre.

TESTING PROCEDURE.—Take 100 ml. of the sample to be examined and mix with it 1 ml. of the orthotolidine reagent, using a 100 ml. Nessler tube. The mixture should be put to stand until the maximum colour has developed, and this standing period should always be at least 5 min. The sample should not be exposed to direct sunlight from the time it is taken to completion of the test, since the sunlight has a photochemical effect and will tend to destroy the residual chlorine. The amount of colour developed will depend on the amount of free chlorine present, smaller amounts giving a yellow colour, while with increasing chlorine the colour becomes more orange. The amount of chlorine present may be quantitatively estimated by comparing the colour produced in the Nessler tube with the colour of a set of standards prepared from the copper sulphate and potassium dichromate solutions as described below. Sometimes with certain waters, such as certain lime-treated or natural waters, a match with the standard tints cannot be obtained; it is then necessary to add more acid.

An important point in using the colour standards listed below is that the colour comparisons be made in 100 ml. Nessler tubes having the graduation marks at 300 mm. from the bottom. Some Nessler tubes have the graduation mark placed 240 mm. from the bottom, and in such cases it is necessary to use other standards developed by Muer and Hale.[2] A variation of 10 mm. is the maximum allowed in the height of the graduation mark in any set of colour comparison tubes used for the determination.

Standards are prepared by mixing the quantities of copper sulphate and potassium dichromate as given in the table and diluting the whole to 100 ml. with distilled water.

TABLE 14

PERMANENT STANDARDS FOR CHLORINE CONTENT

Chlorine, p.p.m.	Solution of copper sulphate ml.	Solution of potassium dichromate ml.
0·01	0·0	0·8
0·02	0·0	2·1
0·03	0·0	3·2
0·04	0·0	4·3
0·05	0·4	5·5
0·06	0·8	6·6
0·07	1·2	7·5
0·08	1·5	8·2
0·09	1·7	9·0
0·10	1·8	10·0
0·20	1·9	20·0
0·25	1·9	25·0
0·30	1·9	30·0
0·35	1·9	34·0
0·40	2·0	38·0
0·50	2·0	45·0
0·60	2·0	51·0
0·70	2·0	58·0
0·80	2·0	63·0
0·90	2·0	67·0
1·00	2·0	72·0

Colour standards for amounts higher than 1·0 p.p.m. are given by Muer & Hale in the reference already cited.

TANNINS

The following method for routine determination of tannins added to feed-water in colloidal treatments or for control of dissolved oxygen was devised by Berk and Schroeder.[3]

SOLUTIONS REQUIRED:

(a) *Tyrosine reagent.*—Prepared by dissolving 100 grams sodium tungstate and 20 grams phosphomolybdic acid in a mixture of 750 ml. of water and 50 ml. of 85 per cent phosphoric acid. The mixture should be boiled under a reflux for 2 hours, then cooled and made up to a litre.

(b) *Saturated sodium carbonate.*—Dissolve not less than 250 grams of sodium carbonate in a litre of water, using heat to effect complete solution. All the solution to stand at 20° C. until crystals of the decahydrate appear, and after standing a further 12–24 hours decant the clear supernatant liquid and store in a bottle with a rubber stopper.

(c) *Master standard solution.*—Dissolve 1 gram of pure tannic acid in a litre of water, and then dilute 25 ml. of this solution to a litre, thus obtaining a final solution containing 25 p.p.m. pure tannic acid.

(d) *Comparison solution.*—Dissolve 5 grams of the tannin extract in use in a litre of water, and dilute 10 ml. of the solution so obtained to a litre. This then gives a solution containing 50 p.p.m. of the extract.

WORKING DETAILS.—Procedure is very simple. A number of standards are prepared from the comparison solution, working on the basis that 1 ml. of this solution made up to the mark in a 50 ml. Nessler tube will give a concentration of 1 p.p.m. Then prepare the unknown by taking sufficient volume of the sample as estimated to contain about 0·15 mgm. of tannin (filtering if the sample is turbid) and also making up to 50 ml. with distilled water in a Nessler tube. Now add 2 ml. of the tyrosine solution to each tube and mix contents carefully, and after 5 min. add 10 ml. of saturated sodium carbonate solution to each tube. Finally, after a further 10 min., compare the blue colour of the sample with that developed by the standards. Simple calculation on the basis of volume of sample taken will then give tannin present in p.p.m.

The comparison solution can be checked by using it as the unknown, and making up the standards from the master standard solution of tannic acid. Various precautions are recommended. Tannins oxidize rapidly in alkaline solutions and therefore the sample drawn for testing should be rapidly cooled and filtered, then made neutral or slightly acid with hydrochloric acid. Any precipitation of tannin experienced in this stage will not matter if the sample is shaken thoroughly before withdrawing a portion for test. Inorganic substances normally present in boiler water are said not to interfere with the determination, though the reaction is not specific for tannins.

REFERENCES

Analysis & Routine Testing

(1) C. Blacher, " Chem. Zeitung," 1912, Vol. 36, p. 541; " Chem. & Ind.," 1912, p. 555; " Chem. Zeitung," 1913, Vol. 37, p. 56; " Chem. & Ind.," 1913, p. 158.
See also recent review of potassium palmitate methods in " Chem. & Ind.," 1935, p. 250T—by P. Hamer.

(2) Muer & Hale, J. Am. Waterworks Assoc., 1925, p. 50.

(3) Berk & Schroeder, Ind. Eng. Chem. Anal. Edn., 1942, p. 456; "Analyst," 1942, p. 372.

APPENDIX

SOLUTIONS

Standard Solutions.

Sulphuric Acid, N/10. Contains 4·9 grams H_2SO_4 per litre of solution, and may be prepared by diluting strong acid to this extent. Standardization, however, is a skilled operation and it is advisable to purchase normal acid and dilute 100 ml. to 1,000 ml.

Caustic Soda, N/10. Contains 4 grams NaOH per litre of solution. Dilute 100 ml. of normal caustic soda to 1,000 ml.

Sodium Carbonate, N/10. Contains 5·3 grams anhydrous Na_2CO_3 per litre. Dilute 100 ml. of normal sodium carbonate to 1,000 ml.

Nitric Acid, N/10. Contains 6·3 grams HNO_3 per litre. Dilute 100 ml. of normal nitric acid to 1,000 ml.

Ammonium Chloride. Dissolve 0·0315 gram of pure salt in a litre of water. (1 ml. then = 0·00001 gram ammonia.)

Wanklyn's Soap Solution. Dissolve 10 grams castile soap in 1 litre of hot dilute alcohol (35 per cent by volume). Cool, filter and test with standard hard water. If necessary, dilute with weak alcohol until 70 ml. of the hard water require 15 ml. to produce a permanent lather.

Clark's Soap Solution. Grind to a fine powder a mixture of lead plaster (Plumbi Emplast. according to British Pharmacopoeia) and anhydrous potassium carbonate, proportions 150:40. Then add a little methylated spirit and work mixture again in the mortar until a smooth cream is obtained. Add more methylated spirit, and mix again; allow to settle and filter. Dilute with aqueous methylated spirits (spirit 2, water 1) until 50 ml. of the standard hard water require 14·25 ml. of this solution to produce a permanent lather.

Standard Hard Water. Dissolve 0·2 gram calcite in dilute hydrochloric acid and evaporate to dryness (in a platinum dish if possible). Add a few ml. of water and evaporate again. Repeat this four or five times to remove the last traces of acid. Then dissolve the residue in water and make up to 1,000 ml. volume.

Potassium Palmitate. Slowly add 25·63 grams of pure palmitic acid to a mixture of 250 grams of glycerine and 400 ml. of alcohol (or methylated spirits) containing 10 per cent of water while warming on a water bath. The mixture must be stirred and kept warm until the acid is completely dissolved. The solution is then cooled, 1 ml. of phenolphthalein added, and titrated with alcoholic caustic potash solution (34 grams KOH per litre) until faint pink colour appears. It is then made up to 1 litre with alcohol (or methylated spirits) containing 10 per cent of water. It may be necessary to add a further few drops of alcoholic caustic potash to neutralize any acidity in the alcohol. The final reagent solution should be neutral.

The reagent should be standardized against barium chloride. 0·611 gram of pure barium chloride dihydrate is dissolved in distilled water and made up to 1 litre. This is a 0·005N

	solution. Take 100 ml. of this and titrate with potassium palmitate as described in the determination of total hardness with the reagent. 5·0 ml. of the solution of palmitate should be required.
Silver Nitrate, N/10.	Dissolve 16·989 grams $AgNO_3$ in a litre of water.

Reagents.

Methyl Orange Indicator.	0·2 gram per litre.
Phenolphthalein.	Dissolved in rectified spirit to 3 per cent solution.
Ammonium Oxalate.	A saturated solution. Solubility at ordinary temperature in water is about 4 grams per 100 ml.
Barium Chloride.	10 per cent solution.
Sodium Oxalate.	Saturated solution contains about 3 per cent salt.
Potassium Chromate. (indicator)	10 per cent solution—must be free from chlorides.
Nessler Solution.	Dissolve 6·25 grams of potassium iodide in about 250 ml. of distilled water, 10 ml. of this solution being set aside, and into the remainder carefully run a cold saturated solution of mercuric chloride. A brilliant red precipitate of mercuric iodide is formed, which at first quickly redissolves on shaking. Addition is continued until a slight permanent opalescence is obtained. Use the 10 ml. of iodide solution set aside to bring back opalescence if too much mercury is added. Then add 150 grams of potassium hydroxide, dissolved in distilled water and allowed to cool, very gradually with continual shaking. The whole is made up to 1 litre and allowed to settle, after which the clear liquid is decanted into a stock bottle.
Phenolsulphonic Acid.	Mix 25 grams of phenol with 370 grams of concentrated sulphuric acid in a round bottom flask. Heat the flask completely immersed in a water bath for six hours. The time of heating may be reduced to 2 hours or less by the substitution of one-third of the above volume of concentrated sulphuric acid with fuming acid, i.e. use two-thirds concentrated acid and one-third fuming acid.
Benzidine Solution.	Weigh out 6·7 grams of benzidine and warm with 500 ml. distilled water to which has been added 20 ml. concentrated hydrochloric acid. Cool, and make up to a litre.

CLARK'S TABLE OF HARDNESSES

(IN DEGREES)

C.C.	0·0	0·1	0·2	0·3	0·4	0·5	0·6	0·7	0·8	0·9
0	..	..	..	..	..	..	..	0·00	0·11	0·22
1	0·33	0·44	0·55	0·67	0·78	0·89	1·00	1·09	1·18	1·27
2	1·36	1·45	1·54	1·64	1·73	1·83	1·93	2·01	2·09	2·18
3	2·28	2·37	2·46	2·55	2·64	2·73	2·82	2·91	3·00	3·10
4	3·20	3·30	3·40	3·50	3·60	3·70	3·80	3·90	4·00	4·10
5	4·20	4·30	4·40	4·50	4·60	4·80	4·80	4·90	5·00	5·10
6	5·20	5·30	5·40	5·50	5·60	5·70	5·80	5·90	6·00	6·10
7	6·20	6·30	6·40	6·50	6·60	6·70	6·80	6·90	7·00	7·11
8	7·21	7·32	7·42	7·53	7·63	7·74	7·84	7·94	8·05	8·16
9	8·26	8·37	8·48	8·58	8·69	8·79	8·90	9·00	9·11	9·21
10	9·32	9·42	9·53	9·63	9·74	9·84	9·95	10·06	10·16	10·28
11	10·39	10·50	10·61	10·72	10·83	10·94	11·05	11·17	11·28	11·39
12	11·50	11·61	11·72	11·83	11·94	12·05	12·17	12·28	12·39	12·50
13	12·61	12·72	12·83	12·94	13·06	13·17	13·28	13·39	13·50	13·61
14	13·72	13·83	13·94	14·06	14·17	14·28	14·39	14·50	14·61	14·72
15	14·83	14·95	15·06	15·18	15·29	15·41	15·53	15·65	15·76	15·88
16	16·00	..	..	..	..	..	..	..	..	..

CHEMICAL FACTORS

Found	Wanted	Factor	Found	Wanted	Factor
CaO-lime	Ca $(OH)_2$	1·321	MgO	2NaOH	1·984
			"	Cl_2	1·759
	$CaCL_2$	1·975	$MgCO_3$	$CaCO_3$	1·187
	$CaSO_4$	2·440	$CaCl_2$	Na_2CO_3	0·955
	$Ca(NO_3)_2$	2·940	$CaSO_4$	"	0·779
	$CaCO_3$	1·785	$Ca(NO_3)_2$	"	0·646
	$Ca(HCO_3)_2$	2·880			
	Na_2CO_3	1·890	$MgCl_2$	"	1·113
	2NaOH	1·427	$MgSO_4$	"	0·880
	Cl_2	1·261			
	MgO	0·719	$Mg(NO_3)_2$	"	0·715
$CaCO_3$	CaO	0·560	$2NH_3$	$(NH_4)_2SO_4$	3·880
	$Ca(OH)_2$	0·740	Na_2CO_3	2NaOH	0·755
	MgO	0·403	"	CaO	0·529
	$MgCO_3$	0·842	Fe_2O_3	$2FeSO_4$	1·902
	H_2SO_4	0·980	"	3CaO	1·054
	2NaOH	0·780	H_2SO_4	Na_2CO_3	1·081
	Na_2CO_3	1·059	"	CaO	0·572
MgO	CaO	1·391	Cl	NaCl	1·650
	$CaCO_3$	2·482	Na_2CO_3	$BaCO_3$	1·862
	$Mg(HCO_3)_2$	3·640	Al_2O_3	$Al_2(SO_4)_3$	3·356
	$MgCO_3$	2·091	"	3CaO	1·650
	$MgCl_2$	2·362	Mn_3O_4	$3MnSO_4$	1·980
	$Mg(NO_3)_2$	3·680	"	3CaO	0·735
	$MgSO_4$	2·985	CO_2	CaO	1·275
	$Mg(OH)_2$	1·447	"	MgO	0·916
	Na_2CO_3	2·629	SiO_2	Na_2O	1·032
N_2O_5	$Ca(NO_3)_2$	1·520	$Na_2Al_2O_4$	Al_2O_3	0·622
	$Mg(NO_3)_2$	1·365			

CONVERSION TABLES

VOLUME:—

1 gallon = 277·5 cubic inches = 0·1605 cubic foot = 4 546 litres.
1 cubic foot = 6·228 gallons = 28·3168 litres.
1 litre = 1·76 pints = 0·0353 cubic foot = 0·22 gallons.
1 cubic metre = 220 gallons = 61,024 cubic inches = 35·3148 cubic feet
1 gallon of water weighs 10 lbs. (1 U.S. gallon of water weighs 8·33 lbs.)
1 cubic foot of water weighs 62·4 lbs.

RATE OF FLOW:—

1 litre per hour = 0·0353 cubic foot per hour.
1 cubic foot per hour = 28·32 litres per hour = 6·23 gallons per hour.

PRESSURE:—

1 lb. per sq. in. = 0·068 atmos. = 2·307 ft. of water.
1 foot of water = 0·4335 lbs. per sq. in.
1 atmosphere = 14·7 lbs. per sq. in. = 33·9 feet of water.

MASS:—

1 kilogram = 2·2046 lbs. = 0·01968 cwt.
1 gram = 0·03512 oz. = 0·0022 lbs.
1 lb. = 0·4536 kilos. = 0·000447 ton.
1 ton = 1016·047 kilos. = 2,240 lbs.
1 U.S. ton = 2,000 lbs. = 907·18 kilos.

DEGREES OF HARDNESS:—

English degrees (Clark's degrees) are grains of calcium carbonate per gallon, or parts per 70,000.

French degrees (Continental degrees) are parts of calcium carbonate per 100,000.

German degrees are parts of calcium oxide (lime) per 100,000.

To convert	*to*	*Multiply by*
Grains per gallon	Parts per 100,000	1·43
Parts per 100,000	Grains per gallon	0·70
English degrees	French degrees	1·43
" "	German degrees	0·80
French degrees	English degrees	0·70
" "	German degrees	0·56
German degrees	English degrees	1·24
" "	French degrees	1·78

Summary of Good Water Conditions

Plant Conditions	Item	Conditions	Remarks
Low and Moderate Pressures (Referring particularly to pressures below 350 lbs. per sq. in.)	pH value	Feed—8·5 Boiler—over 9·5	
	Dissolved oxygen	Feed—under 0·1 ml. per litre	
	Alkalinity	Boiler—not over 50 grains/gallon	This refers to *caustic* alkalinity.
	Total solids	Boiler—not over 500 grains/gallon	Much depends on individual conditions — Lancs. boilers in author's experience have carried up to 4,000 grains per gallon.
	Hardness	Boiler—as low as possible	
	Oil, grease	Complete absence	
	Embrittlement, conditioning ratios.	See pp. 60-64, 113-15, 174, and 184	
High Pressures. 600-850 lbs./sq. ins.	Dissolved oxygen	Feed — less than 0·02 ml. per litre	
	Alkalinity	Boiler — 25 grains per gallon maximum	This is *caustic* alkalinity.
	Total solids	Boiler — 150 grains per gallon maximum	
	Hardness	Zero	
850–1,500 lbs. per sq. ins.	Dissolved oxygen	Feed — less than 0·02 ml. per litre	
	Alkalinity	Boiler — 15 grains per gallon maximum	*Caustic* alkalinity.
	Total solids	Boiler — 100 grains per gallon maximum	
	Hardness	Nil	
	Oil, grease	Complete absence	
	Embrittlement, conditioning ratios	See pp. 60-64, 113-15, 174, and 184	

Test for Scale Deposition

Feed water—determine chlorides and sulphates [say C_F and S_F].

Boiler water—determine chlorides and sulphates [say C_B and S_B].
(blow-down)

Then if:—

$\frac{S_B}{S_F} = \frac{C_B}{C_F}$ . . . No sulphate scale is depositing.

$\frac{S_B}{S_F}$ is greater than $\frac{C_B}{C_F}$ . . . Old sulphate scale is being removed gradually. (In this case, check ratios carefully.)

$\frac{S_B}{S_F}$ is less than $\frac{C_B}{C_F}$ . . . Sulphate is depositing somewhere as scale. (If this deposition is occurring there will be some hardness in the blow-down water, and improved feed treatment is required.)

INDEX

INDEX

D

E

F

T

U

V

W

Z

www.ingramcontent.com/pod-product-compliance
Lightning Source LLC
LaVergne TN
LVHW020629300926
849161LV00004B/653
9780820600987